Becoming a Food Scien

Robert L. Shewfelt

Becoming a Food Scientist

To Graduate School and Beyond

 Springer

Robert L. Shewfelt
Department of Food Science & Technology
Food Process Research & Development Laboratory
University of Georgia
Athens, Georgia
USA

ISBN 978-1-4614-3298-2   ISBN 978-1-4614-3299-9 (eBook)
DOI 10.1007/978-1-4614-3299-9
Springer New York Heidelberg Dordrecht London

Library of Congress Control Number: 2012935517

Printed on acid-free paper

Springer is part of Springer Science+Business Media (www.springer.com)

This book is dedicated to the memory of Dr. Herb Hultin, the best major professor a graduate student could ever have. Stuck in a rickety old van for a 260-mile round trip between Rockport and Amherst MA once a week for a semester with a brash, young, know-it-all graduate student, he taught me about life and what it means to be a scientist. On those journeys we shared our mutual loves of American history and the Red Sox which allowed me to see a side of his quiet, contemplative manner that few of his other students ever saw. He was my mentor, my colleague, my source of funding, and my friend. Herb was a pioneer in food biochemistry helping to bridge the understanding of the role of lipids in edible plant and animal tissue in situ. *He dedicated his life to his science, his students, and most of all his wonderful family. I owe much of any success I have achieved in my profession and many of the thoughts expressed in this book to his guidance and his confidence in my abilities.*

Preface

I was destined to be a food scientist at birth, the son of a food scientist and home economist. My dad completed his Ph.D. at Oregon State when I was a toddler. My mom prepared balanced and nutritious meals for me through my sophomore year at Clemson University when my mom and dad moved away on me. I made up my mind I was going to get a Ph.D. when I was in fourth grade. I thought about majoring in Chemistry when I went to college only to become a confirmed food scientist when I had my first course under Dr. Jack Mitchell. All three of my degrees are in Food Science, and I could not have chosen a better field of study.

This book grew out of my frustrations as a graduate student at the Universities of Florida and Massachusetts. After spending 4 years, mostly as an officer in the US Navy, between my undergraduate years and graduate school, I was ready to continue my education. I viewed graduate school as an extension of my undergraduate studies, but I soon discovered there were qualitative differences between the two. All my professors expected my classmates and me to know the differences, but no one was interested in trying to explain what they were. We all had to learn about those differences on our own. We formed a cadre of students to provide mutual support at Florida. Out of those frustrations I have made a lifetime study of what it takes to be successful in a Food Science graduate program and what it means to be a food scientist.

I started developing the concepts for this book about 35 years ago and have built on them ever since. I have used these ideas in teaching a graduate course, FDST 8110—*Food Research and the Scientific Method*, and an undergraduate course, FDST 4200—*Food Science Forum*. The latter is designed to serve as a bridge from an undergraduate degree and either the food industry or graduate school. I hope that the book will serve as a reservoir of ideas for those beginning a graduate education in food science or beginning a professional career in the field. Although at times it may read as a how-to manual for success in graduate school, it is meant to challenge the reader to study the process, to challenge conventional wisdom, and to develop a career path that maximizes the probability of success both in school and beyond. I have had the opportunity to view food science through the lenses of programs at four universities and service in numerous activities with the Institute of Food

Technologists. This book is thus focused on the field of Food Science, but it may have relevance to other scientific disciplines.

The book would not be possible without the help of all those classmates, professors, students, and colleagues who contributed ideas, comments, and criticisms either during my career or in reading early versions of selected chapters. I refrain from mentioning those as the list would be far too long. The best part of being a professor is the interaction with students as they keep you young, challenge your suppositions, and teach you as much as you can teach them.

Contents

Chapter 1
Research as Process

> *I believe that a graduate student should be able to argue with the major professor. When the student starts winning more arguments than the professor, it is time for that student to graduate.*

> Herb Hultin

I am assuming that my readers are contemplating seeking a degree in food science or are currently in graduate school in Food Science. Current graduate students may be at one of many different stages of development. This book should help in developing as a research scientist. Many students think of graduate school as an advanced form of undergraduate education. I contend that the two are fundamentally different. Most major professors assume that their students know the differences and don't see any reason for any orientation. This book should help students understand those differences and what they need to do to prepare themselves. Here are some of those differences. Graduate education is primarily about

- Research not courses
- Theories not facts
- Questioning not accepting
- Exploring not learning
- Guidelines not authority

This chapter has two objectives: (1) to help prepare one for seeking a graduate program and (2) to get off to a good start in graduate school. The remainder of the book provides things to think about for thriving in, not just surviving, graduate school in food science and to look beyond the degree as a food professional.

I expect undergraduate students to accept the pearls of wisdom I provide in my classroom lectures. I expect graduate students to view what is said in class critically. In graduate school, students should always be questioning what they hear and what they read. They need to develop what we call a thought style (Grinnell, 1992) which we will cover in more detail in Chap. 11. Undergraduate students typically learn by listening, reading, studying, and doing experiments in the laboratory most of which

R.L. Shewfelt, *Becoming a Food Scientist: To Graduate School and Beyond*,
DOI 10.1007/978-1-4614-3299-9_1, © Springer Science+Business Media New York 2012

are directed by others. Graduate students should be exploring knowledge and plotting their own direction. Although graduate students will be pursuing topics of interest to their major professors and the sponsors of their research, they will be expected to be more than sets of hands to perform experiments. It is expected that graduate students will be active contributors to defining the problem, designing the critical experiments and interpreting the data they collect.

Let us begin by thinking about some definitions. We are all scientists in some form of development. We use terms that are critical to who we are, but we might have difficulty in trying to define them. Here are some key definitions selected from Merriam-Webster's (2009) online dictionary (http://www.merriam-webster.com/) to terms that we should know as scientists:

- Science—"knowledge or a system of knowledge covering general truths or the operation of general laws especially as obtained and tested through scientific method"
- Truth—"the property (as of a statement) of being in accord with fact or reality"
- Philosophy— (1) "a search for a general understanding of values and reality by chiefly speculative rather than observational means"; (2) "the sciences and liberal arts exclusive of medicine, law, and theology < a doctor of *philosophy*>"; (3) "the most basic beliefs, concepts, and attitudes of an individual or group"
- Research—"studious inquiry or examination; *especially*: investigation or experimentation aimed at the discovery and interpretation of facts, revision of accepted theories or laws in the light of new facts, or practical application of such new or revised theories or laws"
- Discipline—"a field of study"
- Technology—"the practical application of knowledge especially in a particular area"

Getting closer to home. How can we define food science? Here are three definitions that I have found. Which one is most appropriate for your needs?

- "the application of the basic sciences and engineering to study the fundamental physical, chemical, and biochemical nature of foods and the principles of food processing" (Potter and Hotchkiss, 1999)
- "the physical, chemical and biological properties of foods as they relate to the stability, cost, quality, processing, safety, nutritive value, wholesome-ness, and convenience" (Damodaran et al., 2007)
- The application of physical, biological, and social sciences to the study of food (my working definition)

When entering graduate school, the most important thing expected of us will be to conduct independent research under the direction of a major professor. Course work is required, but it is either to improve our knowledge in a specific area or to expand our knowledge base. We are primarily in graduate school to perform and analyze research, not merely to take courses. As undergraduates, it is important to gain a knowledge base. We talk "facts," but "facts" are primarily for lawyers not scientists.

Life is not as certain for scientists as it is for lawyers. Certainty is a difficult concept to grasp (Burton, 2009). Scientists think in terms of principles, concepts, and theories—not "facts." I consider the term "fact" to be another four-letter word, and I ban its use in my graduate courses and from further use in this book.

In selecting a graduate program, bear in mind that the two most important relationships we will have are with our major professor and with our research topic. We also want to choose an institution with a solid academic reputation (a good football team is an added bonus). The three are intimately related and come as a package deal. We may wish to stay at the school where we are currently studying or we may wish to attend another college. The advantages of staying where we are include familiarization with the school, the local scene, our professors, and their interests. Because of these familiarities it is usually easier to settle into a graduate program, finding research funding if we are well respected and obtaining our degree(s) in a shorter period of time. Bear in mind that the best undergraduate teachers may not be the best major professors in graduate school. The advantages of going to a new school include stretching ourselves and making new friends and contacts that provide more networking opportunities.

Careful selection of our school, professor, and research topic will be critical to success in graduate school and even more important after we graduate. If we have a strong academic record from our previous degree(s), we might be pursued by different potential major professors at our current school even from other institutions. Here are some suggestions for deciding where to go school for a graduate degree(s)

- Start the process at least a year before graduation.
- Try to get some real experience in laboratory research.
- Identify three to four subject areas of research.
- Talk to professors, postdocs, and/or graduate students about possible schools for the types of research of interest.
- Identify three to five schools that appear to be the best possibilities (schools with approved Food Science programs are available at the IFT website – http://www. ift.org/cms/?pid=1000426).
- Search the websites of these departments for faculty members with similar research interests (many departments have a list of faculty research interests like the one in Fig. 1.1).
- Go to the individual page of each professor of interest to get more details.
- Look at a few of the recent publications listed to see if this type of research is really interesting.
- Contact the professors with the most interesting areas of research to see if they might have an opening in their labs (it is generally best not to contact more than two or three faculty members at each institution).

The initial contact with a professor will probably be by e-mail. We need to make sure that our e-mail address is a professional one and our message is professional as well. We may wish to attach a résumé. Also make sure that the résumé is complete, descriptive, well structured, and checked by an advisor and other instructors. Electronic communication is not a substitute for face-to-face meetings. If possible,

DEPARTMENT OF FOOD SCIENCE AND TECHNOLOGY
THE UNIVERSITY OF GEORGIA
Athens, GA 30602 (Phone: 706-542-2286; FAX: 706-542-1050)
http://www.caes.uga.edu/departments/fst

Casimir C. Akoh
Research Professor
Ph.D. Washington State University
cakoh@uga.edu; 706-542-1067

Food chemistry and biochemistry. Chemical and enzymatic synthesis of fat substitutes and structured lipids. Food emulsifiers; enzymatic modification of lipids and phospholipids; synthesis of flavor and fragrance compounds. Recovery of frying oil; nutraceuticals, and phytochemicals.

Jinru Chen
Professor*
Ph.D. University of Guelph
jchen@uga.edu; 770-412-4738

Microbial genetics - rapid detection of bacterial pathogens; epidemiological typing; microbial stress response; bacterial physiology and pathogenicity; elimination of pathogens from food.

Joseph F. Frank
Professor
Ph.D. University of Wisconsin
cmsjoe@uga.edu; 706-542-0994

Dairy and food microbiology; growth and survival of microorganisms in the food processing plant; biofilms; microbial viability detection; dairy fermentations.

Mark A. Harrison
Professor and Graduate Coordinator
Ph.D. University of Tennessee
mahfst@uga.edu; 706-542-1088

Food microbiology and toxicology. Occurrence and survival characteristics of bacterial pathogens in processed food; shelf-life extension of processed food; pathogen detection methodology.

Yao-wen Huang
Professor
Ph.D. University of Georgia
huang@uga.edu; 706-542-1092

Aquatic food technology. Processing and microbiology of fishery and poultry products; new product; shelf-life extension of processed food; by-product recovery and utilization.

Yen-Con Hung
Professor*
Ph.D. University of Minnesota
yhung@uga.edu; 770-414-4739

Physical properties of foods; food quality enhancement; inactivation of pathogens on foods; mathematical and computer modeling of heat and mass transfer; non-destructive quality sensing; postharvest handling of fruits and vegetables.

William C. Hurst
Professor and Outreach Coordinator
Ph.D. Louisiana State University
bhurst@uga.edu; 706-542-0993

Postharvest technology of horticultural crops (fruits, nuts, vegetables). HACCP and SQC (Statistical Quality Control) instruction for fruit/vegetable processing, fresh produce handling, and minimally processed produce.

William L. Kerr
Professor and FPRDL
Coordinator
Ph.D. University of California
wlkerr@uga.edu; 706-542-1085

Physical properties of foods; food processing. Rheological and textural properties of foods. NMR, ultrasound, and calorimetric techniques as process sensors. Computational modeling of food components.

Fanbin Kong
Assistant Professor
Ph.D. Washington State University
fkong@uga.edu; 706-542-7773

Food engineering; physical properties of foods; influence of food matrix and processing on bioaccessibility and bioavailability of bioactives; microencapsulation of bioactive components for controlled release in the human GI tract; membrane filtration.

Ronald B. Pegg
Assistant Professor
Ph.D. Memorial University of Newfoundland
rpegg@uga.edu; 706-542-1099

Functional foods and health aspects of food products.

Anna V. A. Resurreccion
Professor*
Ph.D. University of Georgia
aresurr@uga.edu; 770-412-4736

Consumer preferences. Sensory evaluation. Food quality. Relationship between physico-chemical quality characteristics of raw, processed, packaged and stored food. Modeling and optimization of formulations and processes in food products that utilize plant protein sources. Nutrition.

Robert L. Shewfelt
Meigs Professor & Undergraduate Coordinator
Ph.D. University of Massachusetts
shewfelt@uga.edu; 706-542-5136

Flavor and color quality of foods as evaluated by instrumental techniques, sensory analysis and consumer testing; postharv-est physiology of fresh fruits and vegetables.

Rakesh K. Singh
Professor and Department Head
Ph.D. University of Wisconsin
rsingh@uga.edu; 706-542-1084

Thermal process modeling including aseptic processing and continuous high-pressure, recovery of food processing waste water, and biosensor development.

Louise Wicker
Professor and MFT Coordinator
Ph.D. North Carolina State University
lwicker@uga.edu; 706-542-2574

Protein chemistry, pectin substances, pectic enzymes. Physical properties of foods. Enzymes as process aids. Pectin-protein interactions and colloidal stability of juices, juice drinks, acidified milk drinks, functional beverages. Prediction of performance of ingredients in complex food systems and value added processing of foods for quality, stability and performance.

*Faculty located at:
Dept. of Food Science and Technology
Griffin, Georgia 30223-1797
Phone: 770-412-4758
FAX: 770-412-4748

Fig. 1.1 Examples of faculty research interests for the department of food science and technology at the university of Georgia

arrange a visit to the top two or three schools. Generally these trips are organized through the graduate coordinator's office. Make sure that the professor(s) of interest will be on campus during the visit. If a face-to-face meeting on campus is not possible, meeting potential professors at the IFT annual meeting or other scientific venues is another possibility. Plan ahead. Schedule these events ahead of time via phone or e-mail. Getting someone to help networking at an annual meeting can be a very effective way of meeting people.

When visiting a campus, it is important to realize that we are interviewing each professor just as they are interviewing us. We want to be able to sell our abilities to each professor we meet, but we should be careful in conveying a consistent message. Professors talk to each other and swap stories about students and their approaches. It is also a good idea to talk to some current students in that program to learn more about the faculty. Professors have reputations, some good and some bad. I've known professors who are almost impossible to please, others who are easy to work with but are not particularly good at placing their graduates, and then there are those who push their students hard but the end result is more than worth the extra effort. Never base a decision on one person's opinion.

Sometimes the websites are not keeping up with the latest information, and research interests may have changed. When finding a major professor, we will be working on a current or new project in that lab, usually one that has been funded on a grant. If lucky, we may be given a choice between two or more projects. Likely we will be assigned a project. Before finalizing a bargain with a professor, we should make sure that it is a topic that really interests us. That topic will be a focal point in our life for the next few years. If it does not excite us now when it is fresh and placed in its best light, it will be difficult to maintain interest when immersed in it.

The Major Professor

Major professors are the basis of the graduate mentorship model. It is much like becoming a skilled apprentice to a master developed in medieval times. Major professors operate under their own specific thought styles which will become a basis of our own personal thought style. Major professors provide a ready-made network for help in graduate research, when searching for a job after graduation, and in starting a career. Within this network, for good or ill, we will always be known as Professor X's former student. In some labs, with only a few graduate students and no postdocs, we may have daily access to our major professor. In other labs, we may need to consult with one or more layers of bureaucracy before being able to communicate with our major professor. Regardless, the thought style of the major professor undergirds everything that happens in the lab.

> **RULE # 1**
> The major professor is always right.

Those who don't believe that major professors are always right need to read the preceding paragraph over and over again until they get it! Thus, whenever this book contradicts a major professor, the book is wrong and the major professor is right!

When starting research, we will need to establish a relationship with our major professor. Take a cue from the other students in the lab and carefully observe the protocol in the lab. Some professors have an open-door policy making them available to students at all times they are in the office. Others even make their cell-phone number

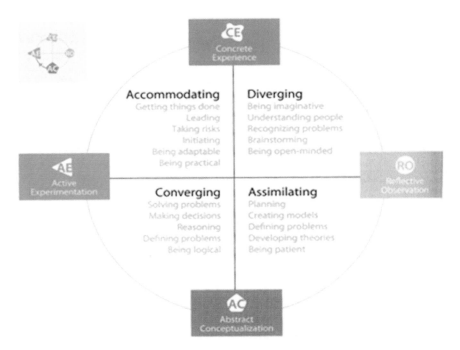

Fig. 1.2 Plot of Kolb's learning style inventory with the strengths of each learning style. *Source*: Kolb learning style inventory. ©2007 Experience based learning systems, Inc. All rights reserved

available to all students or ask to be our friend or colleague on a social or professional networking site. Some will hunt us down when wanting to talk. A few set up barriers to communication. Generally, the earlier in their career and the fewer people in the lab, the more accessible a professor will be. It is important to get help when needing it, but it is also important not to become a pest. When approaching the major professor, we must make sure we have done our homework. Be prepared to answer the most likely questions we will be asked. This is graduate school and professors don't wish to spoon-feed us. Find out who is the best person in the lab who can help. It may or may not be our professor, but it is important that the professor or the lab manager knows where we are in our research and if we are making adequate progress.

When communicating with the major professor, in writing or face-to-face, it is important to understand the professor's perspective. Start by reading a selection of relevant publications, particularly the most recent ones of our professor. A careful reading will help us identify what aspects of the research are important. Review articles and book chapters are particularly useful. Many labs have scheduled group meetings, usually weekly or monthly. Typically, one of the members in the lab makes a presentation of their work and others comment on it. Pay close attention to what excites and dismays the major professor. Part of this perspective is the learning style. There are several different ways of determining learning styles. My favorite as it applies to scientists is Kolb's theory who divides the styles into accommodators, divergers, assimilators, and convergers (Kolb, 1983; HayGroup, 2007) based on the importance we place on experiencing versus thinking and doing versus reflecting (see Fig. 1.2).

Accommodators are better suited for management and entrepreneurs; divergers, for innovation and product development; assimilators, for developing theories and teaching classes; and convergers, for academic research and problem solving. Take the test to learn our preferred learning style and the implications of this type of learner. We can take the test and obtain an analysis for a fee at www.haygroup. com/tl. When we understand our professor better, we might want to take the test from the professor's perspective. If we determine that we have a similar learning style to the professor, we will be able to more effectively communicate with our professor. If we have opposite learning styles, communication will be more difficult, but a combination of different learning styles frequently results in a more complete understanding of the topic. Other ways of viewing learning styles have been described by Gregory (2005) who divides students into visual, auditory and kinesthetic learners. This classification is very useful in teaching students in school education, but is probably not as effective at classifying scientists as the Learning Style Inventory. I will refer to these different types of learning styles in later chapters.

Research as Process

This book grew out of my frustration when conducting my MS research. When starting out in research, students make mistakes. Mistakes and heartache are just part of the experience. Rather than keep my mistakes to myself, I would share them with my classmates who would frequently indicate that they had made the same mistake months before. It seemed that we were all making similar mistakes and failing to learn from each other. I believed that there must be a collective body of knowledge out there that, if tapped, could help students avoid these mistakes and make more "productive" mistakes. It is to these classmates that I dedicate this chapter. I encourage talking with students in the lab and other labs about research experiences. It is through these interactions we can learn more about research and about ourselves.

There are several underlying assumptions I make in this book. Many of these assumptions are controversial. Remember **RULE #1**. Here are my primary assumptions:

- There is no single scientific method.
- Science does not deal with f***s.
- Truth is elusive and is not necessary to make progress.
- Perception is reality to any individual.
- The thought style of the laboratory director (e.g., major professor) governs the scientific progress of that lab.
- Research is a loosely structured, unpredictable process.
- All research fits into a series of unit operations.

Fig. 1.3 Linear diagram of
the unit operations of food
research as composed by
Carlos Margaria

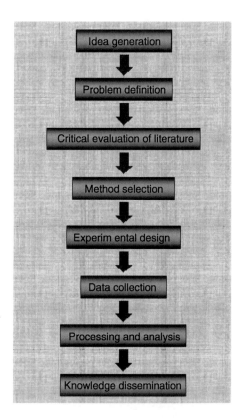

We will come back to each of these points as we progress through the book. Each of the next eight chapters will be devoted to one of the following unit operations of research:

2. Idea generation
3. Problem definition
4. Critical evaluation of the literature
5. Method selection
6. Experimental design
7. Data collection
8. Processing and analysis
9. Knowledge dissemination

We will go into each operation in depth. Although it is easy to think of research as a linear process (Fig. 1.3) it is more likely to be cyclical (Fig. 1.4) and thus never ending. We will approach the process in the order found in the linear diagram. Do not feel bound by convention. You may wish to start with Chap. 9 on knowledge dissemination and work backward to tackle the chapter you feel most comfortable with first or to focus on the operation that is giving you the most difficulty at this point.

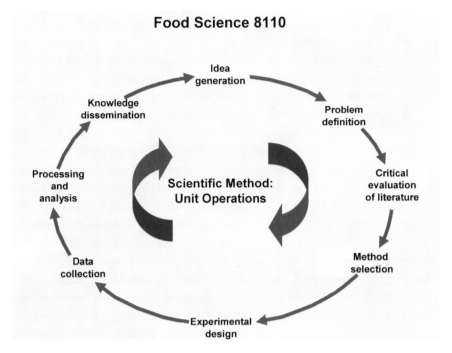

Fig. 1.4 Cyclical diagram of the unit operations of food research as composed by Carlos Margaria

Additional topics that will be covered in the book and may take us beyond the degree we are currently pursuing include

10. The scientific meeting
11. Critical thinking
12. Science and philosophy
13. Ethics
14. Organizing scientific resources
15. Planning
16. Grantsmanship
17. Laboratory setup and management, and
18. Career development

References

Burton R (2009) On being certain: believing you are right even when you are not. St. Martin's Press, New York

Damodaran S, Parkin KL, Fennema OR (2007) Fennema's food chemistry, 4th edn. CRC Press, Boca Raton, FL

Gregory GH (2005) Differentiating instruction with style: aligning teacher and learner intelligences for maximum achievement. Corwin Press, Thousand Oaks, CA

Grinnell F (1992) The scientific attitude, 2nd edn. Guilford Press, New York

HayGroup (2007) Kolb leazzrning style inventory – LSI workbook. Access at www.haygroup. com/tl

Kolb DA (1983) Experiential learning: experience as the source of learning and development. Prentice-Hall, Englewood Cliffs, NJ

Merriam-Webster (2009) Online dictionary http://www.merriam-webster.com/

Potter NN, Hotchkiss JH (1999) Food science. Springer Science + Business Media, New York

Part I
Unit Operations of Research

A Lewis experiment was designed so that a positive outcome would confirm a theory and so that a negative outcome would suggest a pathway for future work. He was not interested in designing an elaborate experiment for its own sake.

Jacob Bigeleisen as quoted by Coffey (2008) about G.N. Lewis, known primarily for Lewis acids but also a key figure in concept development of bond formation in organic chemicals

Reference

Coffey P (2008) Cathedrals of science: the personalities and rivalries that made modern chemistry. Oxford University Press, New York

Chapter 2
Idea Generation

Chance favors only the prepared mind.

Louis Pasteur

Genius is one percent inspiration and ninety-nine percent perspiration.

Thomas Edison

Creativity

Creativity has been defined as

- "the act of generating new and useful ideas, or of re-evaluating or combining old ideas, so as to develop new and useful perspectives in order to satisfy a need" (Quantumiii—http://www.quantum3.co.za/CI%20Glossary.htm)
- "any act, idea, or product that changes an existing domain or that transforms an existing domain into a new one" (Csikszentmihalyi, 1996)
- "purposely making new and valuable products …[to] include significant truths, illuminating explanations, and useful technologies." (Martin, 2007)

A detailed model has been developed by Csikszentmihalyi (1996) who outlines seven steps in the creative process

1. Problem definition and conscious study
2. Focused thinking and unconscious processing
3. "Eureka!" moment
4. Clarification and commitment
5. Experimentation
6. Dissemination
7. Propagation of the idea leading to acceptance

R.L. Shewfelt, *Becoming a Food Scientist: To Graduate School and Beyond*, DOI 10.1007/978-1-4614-3299-9_2, © Springer Science+Business Media New York 2012

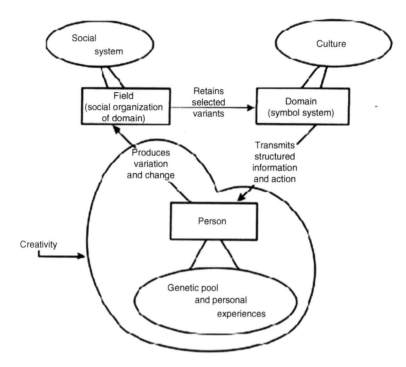

Fig. 2.1 Csikszentmihalyi model of creativity as modified by Weisberg (2006). Reprinted by permission of John Wiley and sons publisher

The first four steps fit into all three definitions shown above, but the last three steps require concrete evidence of creativity. Such a situation raises many questions

- Is someone creative even if what is created is not disseminated?
- Can an idea be creative or must it produce something?
- If an idea is generated today, forgotten, and then revived later and disseminated by someone else, who is creative—the thinker or the disseminator? Read about Mendel in Henig (2001).

The analogy is like the old argument about a tree falling in a forest and whether it makes a sound when it falls if no one is there to hear it. To become a sound, does a person need to hear it or could it be another animal or even an insect? Was Gregor Mendel creative since his ideas were not disseminated until more than 30 years after he completed his research and almost 20 years after his death (see Henig, 2001)? Csikszentmihalyi argues that the creative person must take the idea to a product, but Weisberg has modified Csikszentmihalyi's model to confine creativity to an idea development as shown in Fig. 2.1. Weisberg accounts for the influences of culture (science in our case), the domain (food science), genetics (our innate abilities), and experience (mistakes and insights) on the creative person. The change induced by the creative person could be an ultimate product (new food product, research paper, funded grant proposal) or merely an idea that stimulates creativity for others in the domain.

When starting out my career as a very green faculty member, I had a mentor who may have been the most creative person I have ever met. Any time I went into his office, I would come away with more researchable ideas than I could ever hope to explore. Our small department was incredibly productive, and many of the researchable ideas were directly attributable to him. His publication record was slim, but his ability to generate ideas was particularly impressive. By Csikszentmihalyi's model, he was not creative, but Weisberg would classify him as very creative. However, any creative person must disseminate that information to receive credit leading to Rule#2.

RULE # 2
To obtain credit for any scientific discovery you must be the first person/research group to publish it.

It would appear that there are at least two distinctive types of creativity—breakthrough creativity (Ogle, 2007) and problem-solving creativity (Wakefield, 2003). Breakthrough creativity involves major changes in thinking in an area of research leading to scientific revolutions (Kuhn, 2007) such as the theories of relativity (Einstein, 1920) and the elucidation of the structure of DNA (Watson and Crick, 1953). It tends to favor those who think across disciplines and either ignore some critical theories or are ignorant of them (Ogle, 2007) and appears to favor those individuals or teams who can work across the learning styles described in Chap. 1—assimilation (steps 1 and 2 of Csikszentmihalyi's model), diverging (step 3), accommodating (steps 4 and 7), and converging (steps 5 and 6). Breakthrough ideas also require proper timing and the necessary infrastructure to be accepted and implemented (Ogle, 2007). Creativity does not have to be earth-shattering (Runco, 2003). Creativity is also necessary to solve problems that confront scientists on a daily basis. This type of creativity makes incremental progress pushing the boundaries of accepted theories and principles. It requires the use of critical thinking skills (Chap. 11) and is most effectively employed by a combination of assimilation and converging. I prefer to think of creativity as a continuum ranging from incremental improvement to breakthrough creativity with many intermediate stages between these two extremes. For more insight into the creative process as it relates to scientific discovery, read books by Runco (2003), Simonton (2004), and Martin (2007).

Creativity and productivity appear to be related to age with creativity peaking in the late 1920s and early 1930s for most scientists and downhill from there (Simonton, 2002) Effective scientists are able to combine creativity with experience and resource accumulation to make the greatest contribution in the early 1940s with some variation by field (Fig. 2.2). Productivity can extend up to age 60 (Simonton, 2004).

Fig. 2.2 Contributions to science by age and discipline as plotted by Simonton (2004). Reprinted by permission of Cambridge University Press

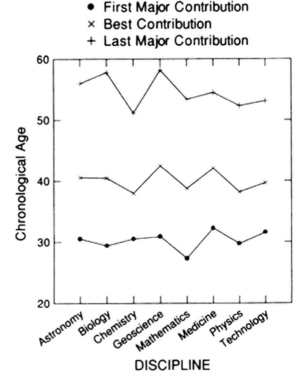

An important concept that goes along with creativity is flow defined as "an almost automatic, effortless, yet highly focused state of consciousness" (Csikszentmihalyi, 2008). Flow is something any creative person must capture to be successful. The characteristics of flow include:

- Setting specific goals
- Obtaining rapid feedback
- Balancing challenges and appropriate skills
- Assessing needs and converting them into action
- Eliminating distractions for complete concentration
- Suppressing any fear of failure
- Losing self-consciousness
- Losing complete track of time
- Developing a cycle of successes

Gough (1952) developed an Adjective Check List to relate to different personality types. He related this list to the creative people (Gough, 1979) and the list is reproduced in Fig. 2.3. Take the test and see how well you score on creativity.

__ affected	__ honest	__ original
__ capable	__ humorous	__ reflective
__ cautious	__ individualistic	__ resourceful
__ clever	__ informal	__ self-confident
__ commonplace	__ insightful	__ sexy
__ confident	__ intelligent	__ sincere
__ conservative	__ interests narrow	__ snobbish
__ conventional	__ interests wide	__ submissive
__ dissatisfied	__ inventive	__ suspicious
__ egotistical	__ mannerly	__ unconventional

Fig. 2.3 Adjective checklist developed by Gough (1952) and evaluated for creativity (Gough, 1979) as referenced by Piirto (2004) and Weisberg (2006). Check all the adjectives that apply. See answers at the end of the chapter

Sociology of Science

Scientists take on many roles in the laboratory (Merton, 1979). Few scientists take on all the roles Merton describes, but most do take on more than one role depending on the situation. They can serve as:

* Technological advisors to graduate students, organizations, federal agencies, and many other groups.
* Technological experts in specific areas of research.
* Technological leaders in that field.
* Sages who are all-knowing persons on a particular topic.
* Scholars who seriously study an area and uncover new knowledge.
* Systematizers who sort information into more understandable forms.
* Experimentalists who publish and contribute to the knowledge base.
* Fighters for truth who argue against myths and legends which become part of popular culture.
* Disseminators of information either in the popular press or in the classroom.
* Creators of knowledge from common problem-solving to development of theories.

What are the preferred learning styles described in Chap. 1 for each of these categories?

When sociologists look at science they see several influences. Society is willing to support science when they see positive benefits coming out of the process. The incredible advances in medicine and treating diseases as well as the successes in space in the 1960s and 1970s have provided science with a good reputation. Space failures, skepticism on global warming, lack of success with fire ants, and highly publicized food poisoning outbreaks have tarnished that reputation. Science follows popular trends, and grant funding calls the shots. Federal and industry dollars are

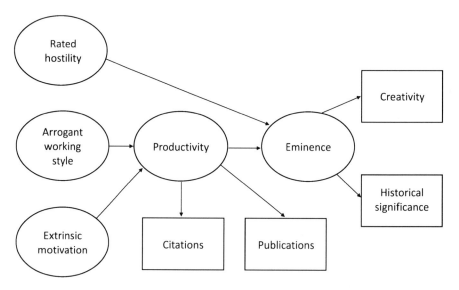

Fig. 2.4 Factors affecting creativity and scientific eminence as modeled by Feist (1993) and adapted by Weisberg (2006). Reprinted by permission of John Wiley and Sons publisher

funding obesity research, the search for "healthy" foods, and food safety. Every 8 years or so there are major shifts in funding brought about by external events and priorities set by the party in power. Recent elections have highlighted both the importance and controversy associated with health care and alternative energy sources.

The reward system for scientists is fairly clear. It is set in numbers of research publications and grant funding amassed. Priority, or the first person/lab that publishes a significant breakthrough, is heralded by other scientists in the field. The "Received" date on the bottom of any research article is the one used to establish priority. Competition for priority among elite scientists is as brutal as for television news scoops. Frequently there are multiple discoveries of key principles due to publication of previous work that does not rise to the level of the big discovery but makes it possible. The most famous multiple discovery is that of Gottfried Leibniz and Isaac Newton inventing calculus (Merton, 1979).

Evaluation of scientists is also a driving force for the scientific enterprise. Most scientists want and seek recognition and excellence. Eminence has been linked to creativity by Feist (1993) and Weisberg (2006) as shown in Fig. 2.4. Recognition comes from salary increases, employers bidding for services, awards, and other recognition. The "Matthew" effect indicates that the first discoverer in a field receives undue recognition while subsequent researchers, even if they have more clear explanations, do not receive adequate recognition (Merton, 1979). In science, the second discoverer is indeed the first loser. I remember hearing the sad story of the second person to independently describe the ethylene pathway in plants with dramatic implications for fruit ripening. He missed priority by two weeks! I heard his story sitting beside him on a bus ride from New Hampshire to Boston, but I can't remember his name. I do know the name of the man who established priority—Shang Fa Yang

(Bradford, 2008). The referee process tends to give the benefit of the doubt to recognized scientists with respect to accepting manuscripts for publication and awarding grants and tends to penalize younger, less recognized scientists.

Welcome to Academe

Promotion and tenure (P&T) is an important process in sorting out ineffective scientists in universities. The typical P&T process brings in a young scientist at the Assistant Professor level. Reputation is established by publications (particularly in prestigious journals), grant funds, invited presentations, teaching, etc. A well-respected researcher who is an adequate teacher is more likely to get promoted than a well-respected instructor who is an adequate researcher. While there are slight distinctions between being promoted and receiving tenure, most scientists promoted to Associate Professor are granted tenure, while tenure is not awarded to those who fail to be promoted within 7 years. For more information on the P&T process, see Chap. 18.

Idea Generation

So where do researchable ideas come from? Anywhere and everywhere. Ideas come from:

- Previous research (a good researcher generates fewer answers than new questions see Chaps. 3, 9, 11 and 12)
- Observation (in daily life, from the news media, from conversations)
- Frustration (things that irritate us both consciously and unconsciously, demanding a solution)
- Funding agencies (use funding as a carrot to study areas they have determined to be important)
- Questions (from annoying people who can't ignore the obvious)
- Dreams and serendipity (weird ideas that just pop into the mind; see Roberts, 1989)

As the quotations that open the chapter indicate, we must be prepared for a good idea when it comes to us and to struggle with it until we can make proper use of it. There are many stories about how ideas were generated. Some of my favorites are:

- Alexander Fleming who saw the future of antibiotics when most would only have seen spoiled plates and a failed experiment (Bankston, 2001).
- Jim Schlatter who noted a sweet taste on his fingers (Robinson and Stern, 1998).
- Friedrich Kekulé who supposedly dreamed of cats chasing their tails that led to proposing the structure of the molecule that is the basis of all phenolic compounds so important in functional foods (Roberts, 1989).

All three provided keys to the important molecular structures shown in Fig. 2.5.

Fig. 2.5 Molecules that
inspired breakthrough
creativity in science.
Can you name them?

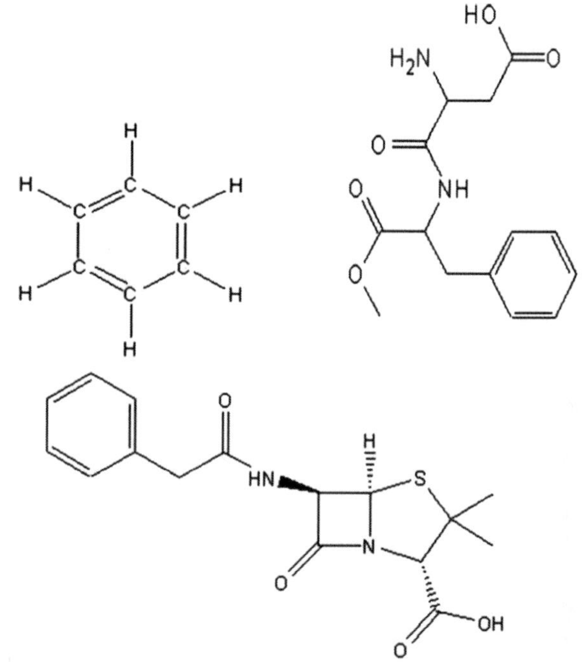

A major portion of my research on the quality of fresh fruits and vegetables over
the past 20 years (Shewfelt, 1986, 1999, 2000; Shewfelt and Prussia, 2009) was
stimulated by an annoying questioning of my major premise by an audacious graduate student who was not even a member of my laboratory (Pendalwar, 1989).

A well-prepared mind belongs to one who reads widely. Such reading includes
popular articles, professional journals, and books. Look for overviews and focus in
on in-depth studies or read about unrelated topics. A successful idea generator is
one who has many ideas and can separate out the really good ideas from the OK
ideas from the really bad ideas. When reading a scientific article hone in on the main
message and then consider the next logical research objective. When evaluating
ideas, ask the following questions:

- Does this idea excite me?
- Can it be formulated into achievable objectives?
- Would these objectives be achievable within a realistic time frame?
- Do I have the capabilities to pursue this idea?
- Do I know someone who can complement my capabilities to pursue this idea?
- Does it have practical significance?
- Will it be viewed favorably by my colleagues and evaluators?
- Is it fundable?

If the answers to enough of these questions are "Yes," then it is an idea worth pursuing. Success is not guaranteed, but our chances for success are better. If there are

several "No" answers to these questions, we must be willing to accept the consequences if we fail. Many successful research pioneers embarked on topics that were not likely to succeed. So too, many who failed to make P&T and sought jobs in other fields. We look more carefully at how to turn an idea into a defined research problem in the next chapter.

Creativity can be cultivated through reading outside our research area. Ogle (2007) says that breakthrough creativity works best when crossing idea spaces. An idea space is similar to a thought collective discussed in Chap. 14. For example, food microbiologists and food engineers operate in different idea spaces with different assumptions and goals. A person or team who can operate in two or more idea spaces can make linkages that suggest new research directions. Ogle suggests that a background in physics combined with limited knowledge in biology allowed Crick and Watson to revolutionize biology by elucidating the structure of DNA. He also indicates that webs of information help create novel ideas that can be exploited.

An example of one chemist whose love of reading led to success is provided by the life of Herbert Brown. He graduated with a degree in organic chemistry, and his girlfriend gave him a graduation present of the only chemistry book she could find, *The Hydrides of Boron and Silicon* (Stock, 1933). Although it had nothing to do with organic chemistry, Herbert was an avid reader, seeing possibilities of working across the idea spaces of inorganic and organic chemistry. He started out his career at the University of Southern California, but he was denied tenure after 9 years. He was able to find a position at Wayne State University and subsequently went to Purdue University. His research in organoborane chemistry was recognized with half of the Nobel Prize in Chemistry in 1979. He and Georg Wittig were recognized "for their development in the use of boron-and phosphorous-containing compounds, respectively, into important reagents in organic synthesis" (http://nobelprize.org/nobel_prizes/chemistry/laureates/1979/index.html). He generously shared his prize money with the girlfriend who had given him the book that started him on his career, which is not that surprising as she had subsequently become his wife (see http://nobelprize.org/nobel_prizes/chemistry/laureates/1979/brown-autobio.html).

Another way of cultivating creativity is by focused thinking. Focused thinking requires complete concentration—no music or other distractions. No multitasking is allowed. A walk in the woods, a comfortable couch in an out-of-the-way venue, lying down in the grass watching the stars, a quiet niche in the library, or daydreaming through an incredibly boring seminar can all be conducive to focused thinking. During focused thinking, we start with a specific or general topic and then let the mind run. There will be diversions to topics that are completely unrelated, but occasional prompting back to the topic at hand may provide some links that are useful. At the end of a focused learning session, a quick debriefing and recording of our ideas for later consideration is advised.

For more ways of cultivating creativity, see suggestions by Piirto (2004).

Answer to Fig. 2.3

Adjectives positively related to creativity:

capable, clever, confident, egotistical, humorous, individualistic, informal,insightful, intelligent, interests wide, inventive, original, reflective, resourceful, self-confident, sexy, snobbish, unconventional

Adjectives negatively related to creativity:

affected, cautious, commonplace, conservative, conventional, dissatisfied, honest, interests narrow, mannerly, sincere, submissive, suspicious

Give yourself a point for every adjective you checked that matches one in the positive attributes and subtract a point for every adjective you checked that matches one in the negative attributes. Top score is +18. Lowest score is −12. How well does this scale really measure creativity?

Answer to Fig. 2.5

Benzene, aspartame, and penicillin.

References

Bankston J (2001) Alexander Fleming and the story of penicillin (unlocking the secrets of science). Mitchell Lane Publishers, Inc., Hockessin, DE

Bradford KJ (2008) Sha Fa Yang: pioneer in plant ethylene biochemistry. Plant Science 175:2–7

Csikszentmihalyi M (1996) Creativity: flow and the psychology of discovery and invention. HarperCollins, New York

Csikszentmihalyi M (2008) Flow: the psychology of optimal experience. HarperCollins, New York

Einstein A (1920) Relativity: the special and general theory. Henry Holt, New York

Feist GJ (1993) A structural model of scientific eminence. Psych Sci 4:366–371

Gough HG (1952) Adjective check list. Consulting Psychologists Press, Palo Alto, CA

Gough HG (1979) A creative personality scale for the adjective check list. J Person Social Psych 37:1398–1405

Henig RM (2001) The monk in the garden: the lost and found genius of Gregor Mendel, the father of genetics. Mariner Books, New York

Kuhn TS (2007) The structure of scientific revolutions, 3rd edn. Univ, Chicago Press, Chicago

Martin MW (2007) Creativity: ethics and excellence in science. Lexington Books, Landham, MD

Merton RK (1979) The sociology of science: theoretical and empirical investigations. University of Chicago Press, Chicago

Ogle R (2007) Smart world: breakthrough creativity and the new science of ideas. Harvard Business School Press, Boston, MA

Pendalwar DS (1989) Modeling the effect of ethylene and temperature on physiological responses of tomatoes stored under controlled atmospheres. University of Georgia, MS thesis

Piirto J (2004) Understanding creativity. Great Potential Press, Scottsdale, AZ

Roberts RM (1989) Serendipity: accidental discoveries in science. John Wiley & Sons, New York

Robinson AG, Stern S (1998) Corporate creativity: how innovation and improvement actually happen. Berrett-Koehler Publishers, San Francisco

Runco MA (2003) Critical creative processes. Hampton Press, Inc., Cresskill, NJ

Shewfelt RL (1986) Postharvest treatment for extending shelf life of fruits and vegetables. Food Technol 40(5):70–72, 74, 76, 78, 80, 89

Shewfelt RL (1999) What is quality? Postharvest Biol Technol 15:197–200

Shewfelt RL (2000) Fruit and vegetable quality. In: Shewfelt RL, Bruckner B (eds) Fruit and vegetable quality: an integrated view. Technomic Press, Lancaster, PA, pp 144–157

Shewfelt RL, Prussia SE (2009) Challenges in handling fresh fruits and vegetables. In: Florkowski WJ, Shewfelt RL, Brueckner B, Prussia SE (eds) Postharvest handling: a systems approach. Academic, San Diego, CA, pp 9–22

Simonton DK (2002) Great psychologists and their times: scientific insights into psychology's history. American Psychological Association, Washington, DC

Simonton DK (2004) Creativity in science: chance, logic, genius and zeitgeist. Cambridge University Press, New York

Stock A (1933) The hydrides of boron and silicon. Cornell Univ. Press, Ithaca, NY

Wakefield JF (2003) The development of creative thinking and critical reflection: lessons from everyday problem finding. In: Runco MA (ed) Critical Creative Processes. Hampton Press, Inc., Cresskill, NJ, pp 253–272

Watson JD, Crick FH (1953) A structure for deoxyribose nucleic acids. Nature 171:737–738

Weisberg RW (2006) Creativity: understanding innovation in problem solving, science, invention, and the arts. John Wiley & Sons, New York

Chapter 3
Problem Definition

Everything should be made as simple as possible, but not simpler.

Albert Einstein

Even though we lump scientific investigation into one word—research, there are many types of research. Some research is directed at achieving a specific goal like finding a cure for cancer or creating the prefect food. Most scientific research is directed at solving specific problems. Most of these problems are narrowly focused. Other research is aimed at developing methods. In food science we may also be looking at optimizing a food process or ingredient formulation. While most food research is applied, basic research seeks a deeper understanding and usually has no immediate application. More will be said on these types of research in the second half of this chapter.

There are many ways to approach problem definition. Mumford et al. (2003) propose four steps in project development from a social science perspective—problem construction, category search, information coding, and category combination. He presents six strategies in problem construction:

- Identify the best possibilities from many alternatives.
- Don't let goal setting get in the way of investigating many ideas.
- Frame the topic from more than one perspective.
- Identify the key components to the overall topic.
- Use analogies from other "idea spaces" to better understand the topic.
- Focus on the main points to prevent information overload.

Spending time developing and evaluating ideas is important. In strategies for further development of ideas and projects Mumford cautions against perfectionism, oversimplification, and goal setting. In the physical sciences, though, goal setting is critical to problem definition. When a problem is framed in the context of a long-range goal, it becomes much easier to plan.

R.L. Shewfelt, *Becoming a Food Scientist: To Graduate School and Beyond*,
DOI 10.1007/978-1-4614-3299-9_3, © Springer Science+Business Media New York 2012

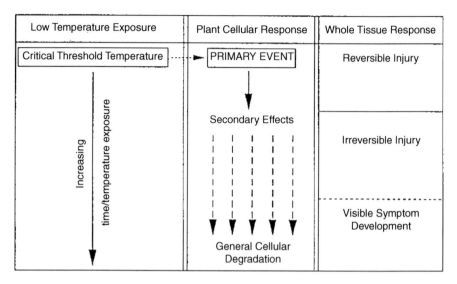

Fig. 3.1 Generalized model for low-temperature injury in the cell and tissue from Shewfelt (1992)

Since coming to my university, I have been involved in three major areas. Although the problems were ambiguous, my research planning began with the development of goals. My first research goal was to develop a systems approach to postharvest handling of fresh fruits and vegetables. This goal grew out of the interdisciplinary effort that was initiated on a research station off the main campus shortly before I arrived and was part of the justification of my position.

We developed a simple model between the field and postharvest system (packing/wholesale/retail) and between the postharvest system and the consumer to frame the boundaries of our system. We demonstrated that the two least understood areas of produce handling were at the boundaries between the field and packing and between retail and the consumer (Shewfelt and Prussia, 1993). Read more about this Postharvest Systems Team in Chap. 15.

My second goal was to better understand the mechanism in the development of chilling injury in susceptible fruit and vegetable species. It grew out of my Ph.D. research with fish muscle as it applied to fruits and vegetables. I developed two models to describe the general perspective at the time (Figs. 3.1 and 3.2). At the time the phase-transition theory of chilling injury (Lyons, 1973) was the best explanation, but there were many doubters (as reviewed by Shewfelt, 1992). Many theories have been expounded since, but no explanation has been generally accepted and no article generated as much research and interest as the phase-transition theory.

My third goal was to relate consumer acceptability of fresh fruits and vegetables to sensory quality and chemical composition. This goal grew out of my experience with the Postharvest Systems Team and led to a quality enhancement model (Fig. 3.3) which was applied to a wider range of products (Shewfelt, 1994).

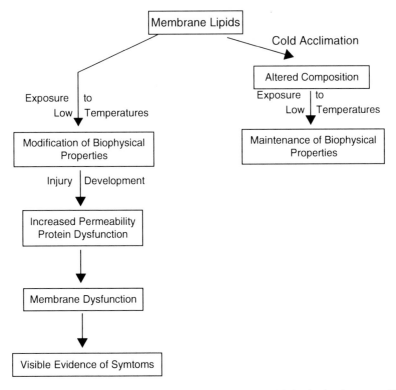

Fig. 3.2 Generalized model to explain the role of membrane lipids in the development of low-temperature injury from Shewfelt (1992)

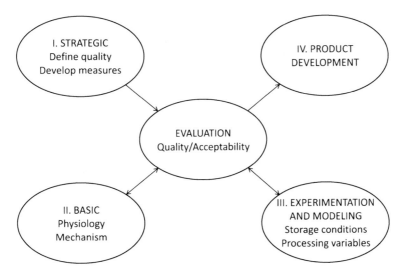

Fig. 3.3 Pictorial diagram of quality enhancement approach from Shewfelt (1994)

Development of a pictorial model helps clarify the problem definition by focusing thoughts and narrowing the scope of study. They may be drawn at the beginning of the research to provide direction, during the investigation to put the work into a wider perspective, or at the end of a project to help explain what was done.

Once a goal has been determined, projects are developed. A project usually is about 2–5 years in duration and has a readily identifiable source of funding. The project objectives are formulated in the context of the long-range goal. Project objectives set the direction to the long-range goal. They should be specific, concise, and to the point. These objectives must yield results that are measurable with accuracy and precision. The objectives must also be achievable within a reasonable time frame, even if the long-range goal is not.

We define a problem to keep a project focused. A clear problem definition also provides a measure of evaluation. It is easy to become distracted and pursue other objectives than those stated. Periodically we need to decide if we are on the right track. If the direction of the project has changed, we need to decide if it is better to return to the stated objective or modify the objective. If we decide to modify the original objective, it may be necessary to consult with the funding company or agency to make sure that the change is appropriate. Clear problem definition also provides a basis for project euthanasia. At times, it is important to decide to stop pursuing a specific research area.

There are times in research where we are faced with what I call the Vietnam Syndrome. During the Vietnam War, opposition grew so large and loud that the US military needed to disengage. Proponents of the war indicated that if we pulled out it would mean that Americans who lost their lives would have died in vain. Opponents of the war indicated that the longer we stayed, the more lives would be lost in a cause that could not be won. We eventually disengaged after losing many more lives. There are times in research we need to determine if we should devote more time and resources into a losing cause. The decision is never an easy one, but it may become moot if all funding dries up. The *Nobel Duel* (Wade, 1981) describes how two colleagues became bitter rivals pursuing a common goal in different labs for over 20 years and ended up sharing a Nobel Prize with someone they despised!

Edward O. Wilson (1998) describes five diagnostic features of research to help us determine whether we are headed in the right direction. They are

- Repeatability
- Economy—minimizing effort to obtain maximum return
- Mensuration—accurate and precise measures
- Heuristics—careful interpretation of the data to set up future experiments
- Consilience—interconnectivity of explanations

Wilson defines science as "the organized systematic enterprise that gathers knowledge about the world and condenses knowledge into testable laws and principles."

Types of Research

Goal-achievement research tends to be general. The long-range goal may be societal such as an end to global warming or the eradication of unsafe foods, basic such as a mechanism for lipid oxidation or ethylene biosynthesis, or applied such as development of a heart-healthy line of food products or the elucidation of tomato flavor. Individual goals may be a small part of a larger goal. This type of research usually must be coordinated with others.

Most research involves problem solving. This type of research tends to be specific. Basic problem-solving research is not tied to a practical application, while applied problem solving is conducted with a practical objective in mind. The problems solved in this type of research may be externally directed by company management, the granting agency, or the thought collective of scientists working in this area who have control over grant funding and research publications. Other problems are defined by individual scientists who then sell the idea to the funding body. Problems may be long term such as minimizing ground-beef outbreaks or short term such as preventing fresh juice from spoiling. The problem may be immediate or part of a broader context.

Another type of research involves methods development. Methods may encompass analytical procedures, processing parameters, or an understanding of systems. Methods research can be a means to an end. Investigators may find that they cannot conduct the necessary research until a specific form of measurement is designed. Other investigators are methods specialists who design new methods as their contribution to science. In general, methods specialists are not considered as prestigious as problem solvers, unless the method(s) they develop revolutionize the field.

Process optimization improves efficiency in food manufacturing, analytical methods, human activity systems, and quality systems. Model building is used to help optimize processes. These models may be physical, mathematical, or theoretical.

A deeper understanding of our world comes from basic research. At it is most basic, it is science for science's sake. The public understands the benefits of highly applied research but may think basic work is meaningless. Applied scientists need a reservoir of basic research to solve practical problems. Very little applied research would be conducted without that basic reservoir. Likewise, companies and governments would be unlikely to fund basic research if they did not derive benefits from applied projects. While frequently considered competitors for scarce research funds, basic and applied scientists work synergistically to advance knowledge. In general, basic scientists tend to have the highest status among scientists.

The dividing line between basic and applied research is frequently blurred. What is typically considered basic research in food science would be considered applied research by a chemist or microbiologist. Louis Pasteur introduced a hybrid of basic and applied research referred to as use-inspired research (Stokes, 1997). Practical problems led Pasteur to plunge in depth to the basic aspects of the problem. Basic research can then lead to different approaches to practical problems.

Defining the Problem

So how do we go about defining a problem—such as the one for a MS thesis or Ph.D. dissertation? We may have been given a clear, concise problem to work on. Chances are that we have only been given a problem area with an ambiguous goal. If so, we should first view it from several angles. Read about the topic in many different sources—including popular, textbook, reviews, and studies. Once we feel reasonably comfortable with the topic clearly described, preferably in 15 words or less, the long-range goal of our research area. It would also be useful to write a similar description of the major goal in the professor's laboratory. Some professors may have one overarching goal, others may be pursuing two or more goals simultaneously. If given a choice, choose the one that relates directly to our topical area. It might be clearly stated in the project proposal for the grant that is funding the research. If working on a funded grant, we might ask to read the project proposal but should approach this question delicately. More secure professors will gladly share their research proposals and appreciate the interest and diligence. More insecure professors will become suspicious that we are trying to steal ideas and could begin to view us as research competitors.

Once discovering the concept of the long-range goal, we need to tease out specific objectives. Again, the project proposal can be invaluable in this exercise. If there is no access to the proposal, recent publications from the lab may point us in the right direction. Stated objectives in a proposal may have been achieved, in progress, ready for study, or modified. To determine the status of these objectives:

- Read recent publications from the laboratory.
- Read recent or forthcoming meeting abstracts from the lab.
- Listen carefully to co-workers in the lab at group meetings.
- Ask co-workers in the lab about their research over lunch or a favorite libation (most of them will gladly share their work complete with hopes, concerns and triumphs if approached in a congenial way).

RULE # 3
Scientific objectives should be clear, brief, achievable and consistent with your goal.

From this investigation, carve out an area that is both interesting and important and write a formal statement of the problem in a sentence of 15 words or less (there is nothing magical about 15 words, but the more clearly and concisely we can state our problem, the easier it will be to sell it to a major professor and graduate committee. Note the Einstein quotation at the beginning of the chapter). From this description, develop the primary objective for our graduate research. Then develop subobjectives needed to accomplish the objective. Each sub-objective will probably become the basis for a chapter in the thesis/dissertation and manuscript to be submitted for publication. Remember that this problem coupled with primary objective and subobjectives are ways to help get us started and not contracts to be fulfilled. Science is rarely so clear that we can state objectives and achieve them without lots of hard work and frequent modification. It is not necessary to consult with the major professor (or postdoc overseeing our project) over every minor modification, but any major modification in the project should be discussed to avoid awkward situations in a final defense!

Defining our graduate problem can serve as preparation for life as an Assistant Professor, should we decide to pursue a career in academe. Here we must hit the ground running with a long-range goal; defined problem; specific, achievable objectives; and fundable grant proposals. Understanding the process on a microlevel will help us negotiate it at the macrolevel in the future. Defining the problem will help us as we proceed to critically evaluate the scientific literature.

Hypothesis Development

A note on the development of hypotheses is needed here. In statistics classes we learn about the null hypothesis and the alternate hypothesis. Exposure to hypotheses may end as we leave the statistics classroom, or they may become a major part of our problem-definition process. Many food science professors insist on us stating an hypothesis. Others have become more cynical because they have found that most hypotheses developed by graduate students are so loosely written and neither testable nor specific enough to be useful. Within the scientific literature, hypotheses are rarely mentioned. Instead the development of objectives and subobjectives that are more open ended and the way science is usually pursued. Doubters should look carefully at the journal articles they read. In the last paragraph of the introduction of almost all articles, there is a stated objective (or aim, the preferred term for British-trained scientists). Another danger of developing a hypothesis is that some scientists fall in love with their hypothesis obscuring some very important insights that their data are trying to tell them (see Chap. 8). Of course, we must not forget Rule #1 or the corollary to it Rule #4.

RULE # 4
Graduation depends on your ability to satisfy your graduate committee.

Chances are that any graduate committee will be composed of members on both sides of the hypothesis divide. For all of the committee meetings, we should be prepared for the question "And what is your hypothesis?" Blank stares or big gulps do not qualify as sufficient answers.

Now a dissatisfactory result on a dissertation defense may not be the end of a scientific career. Svante Arrhenius (Fig. 3.4) did not impress his graduate committee at Uppsala University with his dissertation performance. They indicated that he did not have sufficient data to support his theory. He was given a barely passing grade,

Fig. 3.4 Svante Arrhenius in his laboratory reprinted with permission from Fotosearch. com

but this designation cost him any hope for an academic career. The theoretical ideas in his dissertation provided part of the basis for electrochemistry and contributed to his selection for the Nobel Prize in Chemistry in 1903. Although he was a brilliant scientist who is known for the Arrhenius equation and Arrhenius plots, he apparently never forgave his committee for their failure to fully appreciate his doctoral research (Crawford, 1996; Coffey, 2008).

References

Coffey P (2008) Cathedrals of science: the personalities and rivalries that made modern chemistry. Oxford University Press, New York

Crawford E (1996) Arrhenius: from ionic theory to greenhouse effect. Science History Publications/ USA, Canton, MA

Lyons JE (1973) Chilling injury in plants. Ann Rev Plant Physiol 24:445–466

Mumford MD, Baughman WA, Sager CE (2003) Picking the right material: cognitive processing skills and their role in creative thought. In: Runco MA (ed) Critical creative processes. Hampton Press, Inc, Cresskill, NJ, pp 3–18

Shewfelt RL (1992) Response to chilling and freezing. In: Leshem YY (ed) Plant membranes: a biophysical approach to structure, development and senscesnce. Kluwer Academic Publ, Dordrecht, The Netherlands, pp 192–219

Shewfelt RL (1994) Quality characteristics of fruits and vegetables. In: Minimal processing of foods and process optimization: an interface. CRC Press, Boca Raton, FL, pp 171–187

Shewfelt RL, Prussia SE (1993) Challenges in produce handling. In: Shewfelt RL, Prussia SE (eds) Postharvest handling: a systems approach. Academic, San Diego, CA, pp 27–41

Stokes DF (1997) Pasteur's quadrant. Brookings Institution Press, Washington

Wade N (1981) The nobel duel: two scientists' 21-year race to win the world's most coveted research prize. Anchor, Garden City, NY

Wilson EO (1998) Consilience. Random House, New York

Chapter 4
Critical Evaluation of Literature

> *As the complexity of the world seems to increase at an accelerating rate, there is a greater tendency to become passive absorbers of information, uncritically accepting what is seen and heard.*
>
> Neil Browne and Stuart Keeley (2001)

An effective scientist must develop a balance between the literature and the laboratory. A thorough knowledge of the literature is necessary to learn what research has been conducted in a certain area and identify future needs. Publication of our results must be placed in the context of previous work and point to future directions. This chapter focuses on what to look for and how to evaluate what we find. For details on finding the appropriate literature and how to organize it, see Chap. 14.

Reviewing the Literature

Literature sources can be classified as primary (studies or methods development with original data), secondary (review articles, book chapters, or textbooks), and other aids to finding the appropriate articles to read (collections of abstracts, annotated bibliographies, *Web of Science*). Journals are considered to be prestigious (*Science, Nature*, etc.) when they are highly regarded by scientists in general or are the most respected journal(s) in the field.

Refereed journal articles are those that have been subjected to a peer-review process, typically where two or more experts in the field read the submitted manuscript thoroughly and make recommendations to the journal editor to accept, accept with revisions, or reject the manuscript for publication. In general, the more prestigious journals have a higher rejection rate. Peer review enhances the credibility of the published article. Examples of peer-reviewed journals in food science are shown in Table 4.1. Some book chapters present original data and are peer-reviewed, but

R.L. Shewfelt, *Becoming a Food Scientist: To Graduate School and Beyond*,
DOI 10.1007/978-1-4614-3299-9_4, © Springer Science+Business Media New York 2012

Table 4.1 Partial list of journals that publish peer-reviewed original research articles in food science

Australian Journal of Grape and Wine Research	Journal of Food Biochemistry
Cereal Chemistry	Journal of Food Composition and Analysis
CYTA—Journal of Food	Journal of Food Engineering
Egyptian Journal of Food Science	Journal of Enology and Viticulture
Food Additives and Contaminants	Journal of Food Distribution Research
Food and Agricultural Immunology	Journal of Food Quality
Food and Food Byways	Journal of Food Safety
Food Biotechnology	Journal of Food Process Engineering
Food Chemistry Food Microbiology	Journal of Food Processing and Preservation
Food Quality and Preference	Journal of Food Science
Food Research International	Journal of Food Science Education
Food Security	Journal of Foodservice
Italian Journal of Food Science	Journal of Functional Foods
Innovative Food Science and Emerging Technology	Journal of Muscle Foods
International Dairy Journal	Journal of Sensory Studies
International Journal of Dairy Technology	Journal of Texture Studies
International Journal of Food Engineering	Journal of the AOAC International
International Journal of Food Microbiology	Journal of the Science of Food and Agriculture
International Journal of Food Properties	Journal of Wine Research
International Journal of Food Science and Technology	Meat Science
JAOCS	Lipid Technology
Japanese Journal of Food Science	Lipids
Journal of Agricultural and Food Chemistry	LWT—Food Science and Technology
Journal of Cereal Science	Nutrition and Food Science
Journal of Dairy Science	Quality Assurance and Safety of Crops and Foods

these chapters are usually considered to be of lower status than a peer reviewed journal article. Although most, if not all, of our reading and citations will come from the peer-reviewed literature, there may be other sources of information that will be useful. For example, research reports by agencies, think tanks, and other organizations may provide insight into problems not found in journal articles. Likewise, state-of-the-art articles in trade journals may help in understanding experimental techniques, instrument capabilities, commercial processes, or other commercial applications. Articles in the popular press or on the Internet may also help in framing our problem, but we must be very careful when evaluating them as reliable sources.

When we first start our literature search, we will want to read broadly to get a feel for the field in general. When we are able to clearly define our problem, we will need to be much more selective. If we have only two hours or less a day to read, we need to be very careful in what we decide to read and what we reject. Each paper we read should be directly relevant and applicable to our research.

If we find we do not have enough material to read to occupy two hours of our time each day, we should broaden our reading. When reading in a new research area I start by classifying potential articles on the basis of my reason for reading them (to obtain a general background, help with problem definition, method development, or provide a context for my research). One caution is that we should try to avoid creating what Rorty et al. (2008) refer to as a hermeneutical circle (self-reinforcing evaluation). Hermeneutics is the art of interpretation, particularly as it relates to linking philosophy to research. He warns us that we must look at all of the evidence and not screen out articles and data that disagree with our preconceptions.

Beginning scientists tend to become either overzealous readers or lab enthusiasts leading to imbalances in a research program. Overzealous readers tend to gain a clear concept of the problem and the research done to date but lack the ability to generate data to contribute to the field. Lab enthusiasts can generate large amounts of data, but the experiments frequently reproduce what has already been done or do not address the critical issues in the area.

In screening our reading material, we can proceed through a series of steps.

1. Start with a general background leading to problem definition through textbooks, historical articles, and major reviews proceeding to
2. Methodology papers that provide information on essential procedures, necessary equipment, and supplies that need to be ordered followed by
3. Specific articles to help in narrowing the defined problem through specific results and conclusions as well as current research directions and needs and moving to
4. Finishing touches involving result comparisons, critical evaluation, conception of future studies, and design of specific experiments

This approach works synergistically such that our reading enhances our lab work and our lab research helps clarify our understanding of our reading.

A measure of prestige of a journal is how often the articles published in that journal are cited by other research articles. The impact factor is determined by the Institute for Scientific Information (ISI) and their results are published periodically. The latest information available for Food Science and Technology Journals before publication of this book was in 2011 (http://sciencewatch.com/dr/sci/11/mar27-11_1/). The results for 2005–2009 are shown in Table 4.2. As we can see, review journals are more likely to have a higher impact factor. The reason for this advantage should become clearer in Chap. 9.

In an age where it is so simple to search online, we tend to ignore other important ways to find articles we should read. When searching for keywords in a database, we can run into the Goldilocks Principle in which our list of hits is either too large or too small but rarely just right (Chapin, 2004). Lists that are too large contain many articles that need to be sifted through that are not relevant to our needs. Those lists that are too small can lead us to overlook key articles. Searching for authors of articles we have found useful previously can help us find additional relevant articles. Also, there is a database called *Web of Science* that tracks citations. Thus, we can find recently published articles that cite any key article in our

Table 4.2 Top 10 Food Science and Technology journals from 2005–2009 based on impact factor. The higher the factor the more prestigious the journal http://sciencewatch.com/dr/sci/11/mar27-11_1/

1. Critical Reviews in Food Science & Nutrition (8.96)
2. Trends in Food Science & Technology (7.38)
3. Molecular Nutrition & Food Research (6.82)
4. Food Additives and Contaminants (5.97)
5. International Dairy Journal (5.57)
6. International Journal of Food Microbiology (5.50)
7. Journal of Agricultural and Food Chemistry (5.44)
8. Biotechnology Progress (5.14)
9. Journal of Dairy Science (4.88)
10. Journal of Food Composition and Analysis (4.87)

personal database making it easier for us to keep up with the rapidly growing literature in our field.

While most of our reading should focus on primary articles, key review articles can save us much time in searching through the older literature. If we find some journals that are frequent sources for articles on our list, we can scan the titles in the Table of Contents of key journals as soon as they become available. Also, we should make sure we are keeping current by searching keywords, key authors, and citations of key articles at least once a month. There are ways to receive alerts by e-mail with keywords, authors, and citations of interest. Electronic means help us keep up to date, but we should be careful not to rely on a single method of retrieval.

When keeping up with the literature in our research area, our personal database of articles will accumulate rapidly. A typical MS thesis database should include at least 200 articles with a Ph.D. database exceeding 500 articles. Developing a system to manage this database is critical. For more details on conducting literature searches and organizing our reference database, see Chap. 14.

Reading an Article

"Reading" a scientific article is NOT like reading a novel. It requires concentration and close attention. It may not be as captivating or as entertaining as fiction. It does not need to be read front to back. We should be prepared to take notes. Some scientists work best with electronic copies, others find it easier to scratch up or highlight paper copies. Do whatever works best. Before reading any article, it may help to ask ourselves two questions:

• Why am I reading this article?
• What do I expect to learn from this article?

Table 4.3 Journals that publish primarily review articles in food science

Annual Review of Food Science and Technology
Comprehensive Reviews in Food Science and Food Safety
Critical Reviews in Food Science and Nutrition
Food Reviews International
Food Technology
Reviews in Fisheries Science
Trends in Food Science and Technology

Different types of articles require different reading strategies. We should be focusing on different aspects when reading a review article than when reading a study for the data or reading a methods paper to incorporate a specific procedure.

In taking notes on a review article (see Table 4.3 for journals that publish reviews in food science), we should be able to place the information in the context of a specific aspect of our research. Is it providing insight into a particular area, broadening our horizons, or helping us focus our studies? It is not necessary to read the whole article if there is only one part of it that is particularly relevant to our research. Some review articles have abstracts that might provide context. If not, we should probably read the first and last sections to learn how our interests relate to the interests of the author(s). Skimming the headings for other topics may reveal other relevant sections.

Most studies reporting original data are roughly organized into the following sections:

- *Abstract*
- *Introduction* or *Reviewing the Literature*
- *Materials and Methods*
- *Results*
- *Discussion*
- *Summary and Conclusion*
- *References*

Most journals present the sections in the order above, but some do not. The different sections will be described below in this order, but that is not necessarily the order we need to read them in. Which sections we plan to read and in what order will depend on how we answered the two questions above.

The title of the article is our first indicator as to whether the article is relevant to our research. A well-written title should make it clear if the article is directly related to our work, tangentially related, or unrelated. If it may be related to our work, then we should read the *Abstract*. A well-written *Abstract* has a brief description of:

- The research problem
- The objective of the research
- The experimental treatments and research methods
- The most important results
- The primary conclusion

From this brief overview of the study, it should become apparent whether it is relevant to our research. Life is short. We cannot pursue all the articles that interest us—relevance must take precedence over interest. Sometimes, however, it is the side journeys we take that make the most impact. Be open to new areas, but be careful not to lose sight of our end goal.

If our reading of the *Abstract* indicates that we will read the article and incorporate it into our personal database, the next section we will likely come to is the *Introduction* or *Literature Review*. Relative newcomers to a research area should read this section next to

- Help place the article in context of previous articles
- See what other articles should be read
- Raise new questions not yet considered

The first paragraph in this section usually provides a statement of the research problem in more detail than presented in the *Abstract* and the context for the study being reported. The last paragraph usually states the objective and the rationale for pursuing this objective. Intervening paragraphs provide the chain of articles that lead from the problem to the objective. Some journals permit extensive *Literature Reviews* of up to six or seven paragraphs or even more. Others keep this section short and to the point. If well versed in the literature, we probably want to skip straight to the *Results* section. In this case, the *Introduction* may be the last section we want to read just to scan it for those articles cited that we recognize and those we do not. We may also wish to see if the perspective expressed in this article differs from other articles we have read from this laboratory.

The *Materials and Methods* section is usually not the first section we want to read unless we are reading the article primarily to adapt a method or modify one we are currently using. Adapting or modifying procedures will be covered in greater detail in the next chapter. This section is one we definitely want to read later if the results of this study differ significantly from ones we have read in other articles or from ones we have observed in our laboratory. When reading this section, it is important to determine if the proper procedures were followed as inadequate procedures lead to unreliable results. Apparently contradictory results or conclusions may be easily explained by comparing differences in methodology or experimental treatments. For articles that provide background or context, the *Materials and Methods* section may not be that critical and may be a section we can skip. For any articles that are central to our research, this section must be studied very carefully.

The next section is the *Results* or *Results and Discussion*. A *Results* section generally presents the data in a logical order pointing out the aspects of most interest to the author(s) without explaining the significance of these data. A *Results and Discussion* section usually presents the data from a figure or table and discusses the significance of each before presenting data from the next figure or table. A clearer description of the difference and how we can decide which form to use when writing up our research will be presented in Chap. 9. In this chapter, the two areas are treated as separate sections. Before reading the *Results* section, it can be useful to study the figures and tables. Once we come to this point, we might wish to write down the ideas we get from the data before proceeding to the *Results*. In reading the *Results*, we may find that some of the data in the figures and tables are of much more interest to us than to the author(s). It may be that the authors are presenting data that are directly relevant to our objectives but for a different reason than their objectives. We may also find that the author(s) have pointed out something that makes sense but was not obvious to us when first looking at the data.

Next, we can move to the *Discussion* section in which the author(s) explain the meaning of their data to the readers. Note the areas in which we and the author(s) agree and those which we do not agree. When we do not agree, we should try to find the source of disagreement. Some reasons for disagreement could be

- The author(s) are looking at different aspects than we are
- We have missed an important point or concept or
- There are differences in treatment selection or methodology

The last section in many articles is *Summary and Conclusion*. Sometimes there is just a *Summary* or just a *Conclusion* or they may be wrapped into the end of the *Discussion*. The distinction between the two is that a *Summary* restates the most important points (usually results) made previously in the article, but the *Conclusion* represents a synthesis of what the study means. A *Summary*, by definition, is repetitious. A *Conclusion* is not repetitious.

The section that comes next is the one that most readers skip. Many do not even print out the *References* section when making a hard copy. The *References* can provide some important information. Some questions we might wish to ask, particularly as we become familiar with the topic, are

- Which references are cited but not familiar to us? Do the titles indicate that we should check them out?
- Which references were we expecting to be cited but were not?
- Were the references cited up to date for the time they were published?

Critical Evaluation of Literature

Most scientists are either good experimentalists OR good readers. Highly successful scientists are good experimentalists AND good readers. When evaluating a scientific article, the first glance should include a look at the type (primary, secondary, etc.) of the article, the type of journal (type, refereed, prestige) and the location(s) of the authors. As described above, when we "read" a scientific article we actually need to "study" it using our critical-thinking skills (Chap. 11) to incorporate careful analysis accompanied by some note taking. In evaluating the article, we should ask and be able to answer some, if not all of the following questions

- What are the objectives? Are they stated? Have they been achieved?
- What are the assumptions?—Stated? Underlying? Are they justifiable?

- How were the results displayed and interpreted? Are they related to the stated objective(s)?
- What were the conclusions? Are they relevant? Justifiable? Related to objectives?
- Were the results incorporated into the discussion? Were they placed in context to previous work? What parts of the discussion are supported by the data and what parts are speculative?
- Was the methodology appropriate and repeatable? Are the statistical techniques valid?
- Are the references current, comprehensive, and specific?

Next, we should determine what we can learn and apply directly to our research project.

- How does it relate to previous work? Does it largely substantiate previous research or refute it?
- What new evidence does it provide?
- How specific is this research to meet our needs?
- What new work is suggested? Is that new work relevant to what we are doing?
- How useful would the methodology be to us?

After some experience with the literature, these and similar questions will routinely pop into our minds without prompting. Beginners should, however, develop a set of prompting questions to ensure that they are getting the most out of their reading.

Critical evaluation of the literature does not mean that we reject everything we read, but it does mean that we evaluate each article in context of previous literature. It means that we must resolve any differences in results and conclusions that evolve. It means that we need to be willing to rethink any points where our perspective clashes with what we read or when the articles we are reading now clash with those we have read previously. We should maintain an open mind, but we should not believe everything we read. If we disagree with a point made, we must develop a rational explanation based on evidence, not on mere opinion. Disagreements are frequently the basis for the design of critical experiments that help resolve issues. Allow for the possibility that there is an overriding explanation. Overriding explanations can lead to scientific revolutions when they are right. They can also lead to scientific folly when they are wrong.

Some major advances in food science and technology include the development of canning, freezing, and bulk aseptic processing. Nicholas Appert preserved foods by heating them in a jar leading to the canning process. Peter Durand improved the process by developing the tin can (Marshall Cavendish Corporation, 2007). Francis Bacon is considered the father of scientific experimentation and may have been the first food scientist. He wanted to see if he could preserve chicken by freezing it in the snow. The chicken was preserved, but Bacon died a month later of pneumonia attributed to being out in the cold during the experiment (Marshall Cavendish Corporation, 2007). Clarence Birdseye developed the technology with the help of Donald Tressler and Carl Fellers (Smith, 2002). A method

Fig. 4.1 Dr. Phillip Nelson, Purdue University, World Food Prize Laureate, 2007. Photograph provided by the recipient

to preserve high-quality fruits and vegetables that facilitated storage, packaging, and transportation was designed by Phillip Nelson. He was awarded the World Food Prize for this work in 2007 (http://www.worldfoodprize.org/laureates/Past/2007.htm) and is pictured in Fig. 4.1. The most prestigious award in IFT is named for Nicholas Appert. IFT also presents the Carl R. Fellers award annually to a member who brings "honor and recognition" to the profession (www.ift.org/IFT/Awards/AchievementAwards/).

Not all science results in triumph. An example of folly includes Joseph Priestly's insistence on the phlogiston theory despite evidence to the contrary—primarily evidence he had generated! Phlogiston was a hypothetical component of some materials that allowed them to burn. During burning, phlogiston was postulated to be released into the atmosphere. He discovered a gas that was released from plants that he labeled dephlogisticated air. We call it oxygen. It took Antoine Lavoisier to recognize the importance of oxygen, and he tried to claim credit for the discovery. Priestly is generally recognized as the discoverer of the element, but it probably was Karl Scheele who deserved the credit as Priestly never rejected the phlogiston theory even though his work disproved it (Wengson, 1998).

Throughout this chapter, I have stressed screening articles for relevance to our research as most graduate students tend not to be selective, reading whatever articles cross their computer screen. Some students, however, become too selective and miss key points that are found in closely allied fields. Striking the balance is not easy, but it does become easier as we get further into our research. Finally, it is tempting to ignore all articles that are not written in English or are not available for downloading online. The critical question on whether we should read, study, and incorporate an article into our personal database is the relevance to our research and NOT the ease of obtaining it.

References

Browne MN, Keeley SM (2001) Asking the right questions: a guide to critical thinking, 10th edn. Longman Publishing Group, New York

Chapin PG (2004) Research projects and research proposals: a guide for scientists looking for funding. Cambridge University Press, Cambridge

Marshall Cavendish Corporation (2007) Inventions and inventors, vol 1. Marshall Cavendish, Tarrytown

Rorty R, Williams M, Bromwich D (2008) Philosophy and the mirror of nature: thirtieth anniversary edition. Princeton University Press, Princeton

Smith R (ed) (2002) Inventions and inventors, vol 1. Salem Press, Inc, Pasadena

Wengson R (1998) Scientific blunders: a brief history of how wrong scientists can sometimes be. Carroll & Graf, New York

Chapter 5
Method Selection

> Theorists conduct experiments with their brains. Experimenters
> have to use their hands too. Theorists are thinkers, experimenters
> are craftsmen. The theorist needs no accomplice. The experimenter
> has to master graduate students, cajole machinists, flatter lab
> assistants. The theorist operates in a pristine place free of noise, of
> vibration, of dirt. The experimenter develops an intimacy with
> matter as a sculptor does with clay, battling it, shaping it, and
> engaging it. The theorist invents his companions, as a naive Romeo
> imagined his ideal Juliet. The experimenter's lovers sweat,
> complain, and fart.
>
> James Gleick (2008)

Selection of Appropriate Methodology

Chances are the major professor has a clear idea of what a student project will be
and what methods to use. It may involve the operation of an important piece of laboratory equipment in the lab. Sometimes, however, projects take a turn requiring
development of a specific method to provide specific data. This chapter will cover
some of the things to consider when developing methodology. First let us review the
two previous unit operations—Problem definition (Chap. 3) and critical evaluation
of the literature (Chap. 4).

Before selecting the most appropriate method, it is critical that our problem
is clearly defined. Even before we define that problem we must develop a general background in the research area chosen (i.e., food chemistry, food microbiology, food engineering, nutrition, sensory science, etc.). As we become more
aware of our specific area (i.e., flavor chemistry, pathogenesis, nanotechnol-
ogy), we need to become more narrowly focused in our reading. Further focus
is required when we choose a specific problem or application. As we become
surer of our research topic, we will find that we will need to set limits to read the

R.L. Shewfelt, *Becoming a Food Scientist: To Graduate School and Beyond*,
DOI 10.1007/978-1-4614-3299-9_5, © Springer Science+Business Media New York 2012

Calibration:

1. Press the [stdby] key to start the meter.

2. Immerse electrode in pH 4.0 buffer solution. Stir.

3. Press and release the pH/m/V button until the digital display indicates pH mode.

4. Press the [std] key. After the meter is stabilized, press the enter key. The buffer solution is entered.

5. Thoroughly rinse the electrode with deionized water.

6. Place the electrode in the pH 7.0 buffer solution. Stir.

7. Press the [std] key. Repeat steps 4 and 5.

8. After entering a second buffer, the meter performs a diagnostic check on the electrode. The meter will display "Good Electrode" in the measure mode if the % slope is within 90-102%. If not, it will display "Electrode Error".

Measurement:

1. Always standardize the meter using at least 2 buffers with pH values covering the range of expected pH values of the samples.

2. Make sure the meter is in the Measure mode.

3. Place electrode probe into the sample solution. Stir.

4. When the reading stabilizes, the word "Stable" displays on the screen. Record the pH reading.

5. When finished, return the meter to [stdby] mode. Store electrode in 4M KCl solution. Always leave the filling hole closed on liquid filled units to prevent excessive crystallization inside the probe. Refill as needed when level gets low.

Fig. 5.1 Calibration and pH measurement with the Accumet basic pH meter

most important articles cogent to our research and start generating specific research questions.

In selecting the most appropriate method, we may wish to list all of the methods we require to answer the research question(s) we have developed. These methods should be classified by the analytical techniques that will be employed. At this point, we will want to gather the relevant references for the methods we will need to conduct. Each of these methods can be classified by the basic principle of the method, the required instruments, necessary chemicals, and any special skills needed. In evaluating these methods, we might ask several questions

- Is the equipment available? in our lab? in the Department? somewhere on campus? Have we read the instructions (see Fig. 5.1)?
- How accurate and reliable is the procedure? How accurate and reliable does it need to be to meet our needs?
- How easy is it to perform? How fast does it take to get a particular result? How important is method accuracy and precision relative to the time required obtaining results?
- How important are these data to the success of our project?

What interferes with proper interpretation of the data (e.g., a compound in our samples that interfere with a color reaction or an inhibitor of a specific enzyme)? In selecting the best method, we should consider the theory of each possible method, advantages and disadvantages of the methods mentioned in the scientific literature, and personal experiences of those in the lab who have used these methods. Chances are that someone in the lab with the equipment we need knows this information or someone who knows.

Laboratory Adaptation and Research Plan Development

Once we have selected our method, we must adapt it to our laboratory setting. Methods look so easy to perform when we read about them in a scientific article, but they prove much more challenging when we actually need to perform them. What I have found helpful is to write them out in recipe form. I generally start with a list of all the materials I will need to perform the experiment and then write out a step-by-step procedure. At this point, we need to allow for modifications as we may not have the same equipment described in the article, but we may have something that will serve as an acceptable substitute.

The first trial does not usually work as planned. Do not expect it to work. Pay attention to what is needed. I recommend a walk-through to see that we have all that is required before we try it for the first time. Before starting, check on the equipment availability. There may be a sign-up sheet to schedule use. If training on a piece of equipment such as the one pictured in Fig. 5.2 is necessary, arrange for the training session at the convenience of the trainer. Then gather all of the needed materials and prepare any solutions needed. Make sure these solutions and materials are stored properly and will not expire before needed. Then develop a plan that will avoid having to do two things at the same time or have someone available who can help when things get tense. Some steps may be time sensitive. When these steps occur, we must be ready or jeopardize losing the whole experiment including our time and materials.

In the early trials, we are developing the method and our familiarity with it. I had a colleague who was brilliant at starting small and building. He was doing enzyme assays. His first few times he started with two 25-ml Erlenmeyer flasks in a shaking heated water bath, expanded to four, then eight and up to thirty-two. He staggered each flask by 20 seconds, which meant that he had something to do every 20 seconds over a two-hour period. Any disruption would crash the whole experiment. He was a master at performing these experiments and could generate large amounts of high-quality data in a relatively short period of time. It is important to establish our limits based on equipment availability and our ability to concentrate and avoid disruption. Remember that we are not saving time if we ruin experiment after experiment. The book by Jeffery Mayer (1991), *If you haven't got the time to do it right, when will you find the time to do it over?*, seems appropriate to adapting our methods to our situation.

Fig. 5.2 Gas chromatograph/ mass spectrometer in the laboratory of Dr. Stanley Kays. Photograph by Danielle Wedral

We should know reasons for each step. If we need to make a modification of any procedure, we must modify it as necessary within constraints of theory. Why is each step necessary? I once had a procedure with a step that required me to wait 15 minutes without any specific reason. I tried eliminating the step, but it led to inconsistent data, so I put the 15-minute wait back into the procedure. Once we have a method we can live with, write it out in detail, draw a schematic (for example, see Figs. 5.3 and 5.4) and stick with it. I heard of a laboratory technician who never performed a procedure the same way every time he tried it. Such a practice makes it impossible to compare the data from experiment to experiment.

Any method selected should be part of an overall research plan. Our research plan should incorporate our research objectives and required procedures. We need

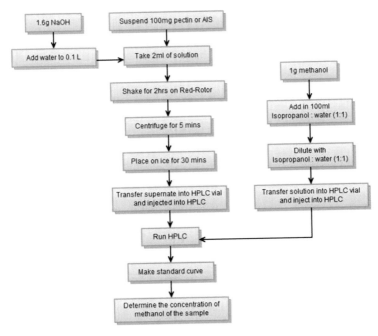

Fig. 5.3 Example of a method schematic prepared by Xiomeng Wu

to develop a work schedule that incorporates equipment availability, a sampling schedule, procedure limitations, material limitations, and time limitations. Our plan should also include a timeline.

> **RULE # 5**
> Pilson's Law "It always takes longer."

Pilson's law (Pilson, 1980) is universally applicable, particularly for graduate projects. Plan in some contingency time as something is likely to happen that will cause delays, particularly when depending on a piece of equipment that could become inoperable and someone else to help or read the thesis or dissertation. If we schedule our time too closely, anything unexpected will jeopardize our time goals. Having said this, we should not ever feel comfortable because we are ahead of our schedule. In our plan, we need to make sure we have allowed for data collection and analysis. Finally, before we get too deep into our experiments, we should obtain feedback on the plan and be prepared to make modifications as necessary.

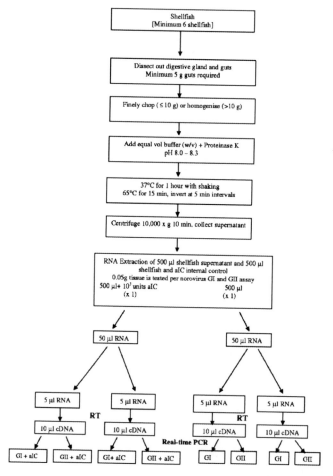

Fig. 5.4 Example of a method schematic form Greening and Hewitt (2008). Reprinted with permission from Kluwer Academic Publishers

References

Gleick J (2008) Chaos: making a new science. Penguin Press, New York

Greening GE, Hewitt J (2008) Norovirus detection in shellfish using a rapid, sensitive virus recovery and real-time RT-PCR detection protocol. Food Analyt Method 1:109–118

Mayer JJ (1991) If you haven't got the time to do it right, when will you find the time to do it over? Fireside, New York

Pilson MEQ (1980) Pilson's law. J Irreproducible Results 26(1):15–16

Chapter 6
Experimental Design

And all this science,
I don't understand.
It's just my job, 5 days a week,
a rocket man.

Elton John and Bernie Taupin (1972)

There are lies, damn lies and statistics.

Attributed to many sources including Benjamin Disraeli, Mark
Twain and Fiordello LaGuardia

Statistical Methods

Statistics can be manipulated by other professions; it is critical that scientists follow strict rules in the use and application of statistical techniques. The use of statistics becomes our referee to help us decide if our ideas and suppositions are correct. If we design our experiments intelligently, collect our data accurately and analyze them correctly, we can determine if our experimental treatments produce clear effects. Statistical analysis does not provide 100% certainty, but it does provide us an objective basis to draw conclusions based on recognized techniques and accepted guidelines rather than mere hype or speculation. A statistically significant difference does not necessarily mean that the treatment will have a practical effect. For example, a small, but statistically significant color change may or may not affect consumer acceptability of a chocolate pudding as the typical consumer may not be as sensitive to color differences as a colorimeter or trained sensory panel. Likewise, if no statistical significance is found in the development of an off-flavor, we conclude that there is no significant difference in the experimental treatment. There may be consumers that can detect the specific off-flavor, but the general population shows no effect.

This book is not a statistics book. There are many fine books out there that can guide us to the proper design of our experiments such as Cox and Reid (2000), Mason et al. (2003) and Urdan (2010). This chapter will outline some basic statistical

R.L. Shewfelt, *Becoming a Food Scientist: To Graduate School and Beyond*,
DOI 10.1007/978-1-4614-3299-9_6, © Springer Science+Business Media New York 2012

terms principles that every graduate student in the field should know. For greater detail, go to the statistics book used at your university or, better yet, the most up-to-date version of the one on the bookshelf of the major professor. Ideally, we would like to design experiments that give us a yes or no answer (Beveridge, 2004). These experiments are possible and are found in some mathematical, physics, and chemistry problems. For example, the combination of two chemical reagents either results in a reaction or they do not. Biological and food systems, however, are not as clear-cut and require a means of deciding whether an observed effect is significant or not.

Statistical probability is a useful measure of that significance. Statistical analysis is a tool to evaluate the significance of our data. It is critical that we choose our statistical methods BEFORE we conduct our experiments. Our experimental design and statistical method are inextricably linked. We should have a reasonably clear idea of what we want to do BEFORE consulting a statistician. We should listen carefully to the statistician, study the suggestions, and then question everything we do not understand. This chapter should help in assessing our situation and formulating our suggestions before we visit with our statistician. We might also wish to consult with our major professor and other members of our committee before putting our plan into action. First, we should familiarize ourselves with some terms. Statistical analysis of data will be covered in Chap. 8.

Common Statistical Terms

Some common terms that many students do not seem to completely grasp include:

Bias is introduction of a systematic error into an experiment either intentionally or unintentionally.

Dependent variables are the measurements made by the researcher that may vary as the *independent variables* are changed. *Dependent variables* are plotted on the *y*-axis in a two-dimensional plot.

Error relates to an estimation of the degree of uncertainty of a particular value, frequently expressed as the *standard deviation* or *standard error of the mean*.

Experimental design is the statistical organization of the study to include the plan to collect and analyze the data. The *design* should be selected BEFORE the data are collected.

Factorial experiments are analyzed for the *interaction* and *main effects* of the treatments chosen by the researcher.

Factors are independent variables such as time, temperature, and relative humidity in a storage study.

Independent variables are chosen by the researcher to investigate and are plotted on the *x*-axis in a two-dimensional plot.

Interaction effects are the influence of one factor on another factor, such as time and temperature. A *main effect* cannot be analyzed in and of itself before taking into consideration the interaction effect.

Levels are the specific values of each factor being studied such as 10, 20, 30, and 40°C or 3, 6, 9, and 12 months in a storage study.

Normal distribution describes a dataset that forms a bell-shaped curve around the *mean*. In most statistical methods we assume a normal distribution, but some of our datasets are NOT normally distributed.

Probability (*p*) is the likelihood that the effect is due to the experimental conditions and not to pure chance. By convention, we usually accept a difference at $p < 0.05$ as significant which provides a confidence of 95% that the effect is real. We also tend to consider a $p < 0.01$ as highly significant (99% level of confidence).

Randomization is a selection process that provides the opportunity of every point in the population has an equal chance of being selected in a sample. It also sets the order of the samples selected.

Replication is the repetition of a measurement. Pipetting three samples out of the same flask does NOT provide three replicates. Conducting an experiment three times under the same conditions can provide replicates (or triplicate samples).

Sample mean is the numerical average of all values in a sample which we assume is representative of the numerical average of all values in the population.

Sampling is collection of a selected subset of the entire dataset (the *population*) which must be *random* to minimize *bias*.

Treatments are the combination of factors and levels. If an experiment has three levels each of three factors, it has 9 treatments.

Trends are predictable directions of a series of data. Trends should be statistically significant and can generally be determined by *regression* techniques. Many students incorrectly use the word trend when they see a pattern in their data that their statistical analysis cannot verify.

Type-one error concludes that there is a difference between treatments when there is no difference.

Type-two error concludes there is no difference between treatments when there is a difference.

For more details on these and other statistical terms, see http://www.stats.gla.ac.uk/steps/glossary/ or the glossary in any other common statistics text such as Urdan (2010).

Table 6.1 shows an unlabeled dataset from a factorial experiment. Look over the data carefully. We should be able to answer the following questions:

- How many factors are there?
- How many levels are there for each factor?
- How many total treatments?
- How many replications?
- Is the experiment randomized? Is there a bias in the design?
- Do the arrangement of the data and number of samples for each treatment provide any clue about how the data were collected?
- How precise are the measurements?
- Which of the methods described below would you use to analyze the data?

Table 6.1 Can you identify the key components of this dataset?

0	0	0	6.5	5.0	5.5	6.0	6.5	7.0	6.5
0	1	1	5.5	5.0	7.0	6.5	6.5	7.5	6.0
0	0	2	4.5	5.5	6.0	6.0	5.0	7.0	7.0
1	1	0	5.5	5.0	5.5	5.5	7.0	7.0	5.0
1	0	1	6.0	6.0	6.5	6.0	6.5	6.0	7.5
1	1	2	5.5	5.5	6.0	7.5	6.0	6.0	8.0
1	0	0	6.0	6.0	6.5	6.0	4.0	4.5	8.0
1	1	1	6.0	7.0	7.0	7.0	6.0	7.0	6.0
1	0	2	6.0	6.5	5.0	6.5	6.0	4.5	7.0
0	1	0	6.0	7.5	6.0	7.5	7.5	7.0	6.5
0	0	1	6.0	6.0	5.0	6.0	6.0	7.0	7.5
0	1	2	6.0	7.0	7.0	5.0	7.5	7.0	6.0

Some of the details of experimental design and answers to these questions will be revealed at the end of the chapter.

Common Statistical Techniques

There are many types of statistical tests. Choosing the right test to analyze our data is critical. These selections should be done in consultation with the major professor, advisory committee, and a statistician. If we use statistics extensively or are using statistical techniques unique to our chosen topic, we should have a statistician on our advisory committee. Generally, we choose the simplest test available that will give a meaningful answer to the research questions posed. Three of the most common techniques include the following:

A *t-test* is conducted when comparing two treatments. It calls for a very simple experiment, but if we want to know which of the two major colas or peanut butters is best, incorporation of any other treatments will just add clutter.

ANOVA (*Analysis of Variance*) is one of the most frequently used statistical techniques. It is ideally suited for *multifactor* experiments and uses the variability in the data to determine the *p-value*. It is particularly effective at drawing conclusions about *interaction effects*, but these effects might obscure main effects of a specific factor. For example, I designed several complex experiments on the changes in quality of fresh vegetables from the field to the consumer. As we were analyzing our data, one of my collaborators would always want to know the *main effect* of temperature of handling and storage without the interfering *interaction effects* of time and harvesting factors.

Our choice of experimental design is directly linked to the selection of our statistical technique. Examples of sample design include *Latin-square, randomized block*, or *split-plot designs*. The power of our analysis and credibility of our conclusions can be affected by our selection of the (in)appropriate design. There are other techniques such as *correlation, regression,* and *means separation* which are applied to the data in conjunction with analysis of variance. These and other techniques and the dangers of misapplying them are described in Chap. 8.

When meeting with the statistician make sure to clearly state the research objective and identify logistical limitations. The statistician will probably try to obtain the full power out of our analysis, suggesting a very ambitious regime that might require our ten closest friends five years to complete. In addition, the statistician may recommend impractical trials. For example, I was once doing a simple storage test at three temperatures (5, 20, and 35°C). The statistician wanted me to run the experiment three times changing the temperature in each room each time I ran the experiment. My experiments were not the only samples in those storage rooms. It would have made at least five other investigators irate if I tried to modify the temperature in each room. In addition, the temperature controls on each room were not able to reliably meet the other temperatures such that the refrigerator (5°C) could not be turned into an incubator (35°C). When the desires of the statistician cannot be met, we need to negotiate such that our results will be valid but under specific constraints. For a more in-depth discussion on how to understand experimental design, I suggest Cox and Reid (2000) or Mason et al. (2003). For a lighter approach look at Best (2001).

Answers to Questions Raised in Table 6.1

It is obvious that there are three factors as represented in the first three columns with a total of 12 treatments ($2 \times 2 \times 3$) in a balanced design. The next seven columns are responses to the treatments. Seven responses suggest that they represent days of the week. If that is so, day of the week would also be another factor, making it 84 treatments ($2 \times 2 \times 3 \times 7$). With 84 treatments, there is no replication. All responses are rounded off to the nearest 0.5 and the range is 4.5 to 8.0. To be analyzed statistically, the entire experiment would need to be replicated.

References

Best J (2001) Damned lies and statistics: untangling numbers from the media, politicians and activists. University of California Press, Berkeley

Beveridge WIB (2004) The art of scientific investigation. Blackburn Press, Caldwell

Cox DR, Reid N (2000) The theory of the design of experiments. CRC Press, Boca Raton

John E, Taupin B (1972) Rocket Man (I think it's gonna be a long, long time). In: *Honky Château.* Universal City Records, Santa Monica. http://www.berniejtaupin.com/discography.bt?ds_id=302

Mason RL, Gunst RF, Hess JL (2003) Statistical design and analysis of experiments, with applications to engineering and science. Wiley, Hoboken

Urdan TC (2010) Statistics in plain english, 3rd edn. Lawrence Erlbaum Associates, Mahwah

Chapter 7
Data Collection

I need to take some samples.

Dr. Grace Augustine shortly before her untimely death in
Avatar

The primary mission of a graduate student is to collect data.

Every step performed before starting to collect data is to prepare the way to collect data. To review, an idea we generate is formulated into a testable problem. The literature we read is selected to illuminate our understanding of that problem and then read in context of the problem. Our reading may modify our idea and force a reformulation of the problem. Methods are selected on the basis of the criteria we develop to test our idea and experiments are designed to give us an answer.

Every step after collecting data is to properly interpret it and to prepare us for the next phase of data collection. The data collected are processed and analyzed to see if the results obtained were expected or surprising. The results may need to be verified or may suggest new experiments. Once a comprehensible body of knowledge is obtained it is time to write an abstract for a conference presentation or preparation of a manuscript for a journal article, thesis or dissertation. Every good research project stimulates more ideas and research questions, which then continues the cycle.

Before proceeding further, it is important to clarify some terms and their usage. For example, "data" is defined as "factual information (as measurements or statistics) used as a basis for reasoning, discussion, or calculation" and is the plural form of "datum." (Merriam-Webster's, 2009) Common usage now permits the use of "data" as either a singular collective noun (the data is) or as a plural noun (the data are). If we run into someone who is old school, we should use the plural form as is used in this chapter and the rest of the book.

Although we tend to rely on scientific instruments, there are many other ways to collect scientific data. Typically one or more instruments in the lab of a major professor keep that lab going. They are the basis for the manuscripts coming out of the lab and a key component in the grant proposals needed to support the lab and its

R.L. Shewfelt, *Becoming a Food Scientist: To Graduate School and Beyond*,
DOI 10.1007/978-1-4614-3299-9_7, © Springer Science+Business Media New York 2012

members. These pieces of equipment are very expensive to buy, to maintain, and to repair. They must be handled with care. The instruments can provide us with the data we need, but an equipment failure can destroy our research and deal a serious blow to a major professor's work.

It is important to learn how to operate the equipment we need in our research. Read the manual. Listen carefully to the person conducting training on how to use the equipment. Do not worry about asking stupid questions. If something is not clear, continue to ask questions until it makes sense. Make sure to learn the proper way to start and shut down any piece of equipment. Also, it is important that we understand the basic principles of the instrument so we can properly interpret the data. If at any time the instrument is not performing as it should, contact the proper person.

> **RULE # 6**
> Don't continue to use a piece of equipment if you think it is not operating properly.

It is also critical to plan ahead for compiling our data. Most data today are maintained in electronic databases. Recognizing how important our data are, we should have them in at least two locations. Maintenance of our data on a computer and an external drive is acceptable, but we may wish to periodically print hard copies if our data is in a form that hard copies will suffice. Organization is a key to any data collection operation. Make sure that all data are assigned with a particular date and a particular procedure used to collect that data. If there is any special sample information, unusual circumstances (e.g., a power outage that delayed reading of some samples), or unusual observations (e.g., a solution turned red when it normally turns blue), make sure to have some indication. If we wish to exclude data or provide alternate explanations, we must have a legitimate reason. When the data are collected, it may be obvious what happened, but when trying to reconstruct the data for a manuscript or grant proposal, the details may not always be as clear.

Data Collection Forms

Once upon a time there was no electronic data collection. The world was much simpler then. Now there are many ways to collect and amass data. Before any data are collected, it is important to clearly visualize how the data will be collected. Some questions include:

- Will they be in electronic form?

  - Do they need to be transferred to a spreadsheet?
  - Will anything be lost during such a transfer operation?
  - Is the spreadsheet compatible with the statistical analysis program?

- Do we need to prepare forms?

    - Are the forms providing all the needed information?
    - Does the written procedure (Chap. 5) guarantee collection of all necessary data? If not, the procedure needs to be modified.

Careful attention to these points can help identify weaknesses or gaps in our plan thus saving us from much grief when data collection begins (see Table 7.1). Times during the data collection phase can be hectic and not allow for reflection or correction. It is a time where we must ensure that the quality of our data is as good as it can be. It is our responsibility that the procedure is adhered to faithfully such that results from one session can be compared to all other sessions. A complete focus may be required for data collection to be successful. I have been told that when I am in a data collection mode, I turn into a completely different person shutting out everything but my mission. We need to be on the lookout to observe any problems or difficulties that may affect our results. In data collection, I rely on Rule #7.

> **RULE # 7**
> "Don't think. Just pitch"—Kevin Costner in *Bull Durham*

We must do all of our thinking ahead of time as the middle of an experiment is not the time to make major changes in our procedure. This rule does not relieve us of heightened attention and concentration on details. One of my colleagues in graduate school noticed one day that all of her data were much lower than that collected on previous days. Upon completion of her readings on the spectrophotometer, she went back through her procedures in her mind to see if she had made a critical error. One possibility was that the spectrophotometer was set on the wrong wavelength. That supposition turned out to be correct. Fortunately she had not discarded any of her samples, and the reaction was not time sensitive. She reread all of her samples and saved four hours of her time to conduct the experiment as well as all of the materials she used.

> *For example, if you are doing an experiment, you should report everything that you think might make it invalid—not only what you think is right about it: other causes that could possibly explain your results; and things you thought of that you've eliminated by some other experiment, and how they worked—to make sure the other fellow can tell that they have been eliminated.*
>
> Richard Feynman

Table 7.1 Excel data sheets for banana storage test. Sheet A outlines physical measurements with calculations for the peel size (column I) and pulp to peel ratio (column J). Sheet B outlines the values for sensory descriptors in the same study. All treatments and calculations are entered prior to conducting the study. Careful setup of data sheets before starting a study helps ensure that all tests are set up correctly

Sheet A

	A	B	C	D	E	F	G	H	I	J
1	Date	Drawer	Finger	Weight	Finger diameter	Pulp diameter	Shear	Brix	Peel	Pulp/peel
2	19-Jan	91	B						=E2-F2	=F2/I2
3	19-Jan	3	B						=E3-F3	=F3/I3
4	19-Jan	11	B						=E4-F4	=F4/I4
5	19-Jan	101	B						=E5-F5	=F5/I5
6	19-Jan	89	B						=E6-F6	=F6/I6

Sheet B

	A	B	C	D	E	F	G	H	I
1	Date	Panelist	Sensory code	Sweet	Green	Banana	Off-flavor	Firm	Juicy
2	19-Jan	A	264						
3	19-Jan	A	616						
4	19-Jan	A	333						
5	19-Jan	A	211						
6	19-Jan	A	720						
7	19-Jan	A	136						

The Laboratory Notebook

In earlier times when many older major professors now were graduate students or postdocs, a hand-written laboratory notebook was required in most labs. It was typically one of these hard-bound notebooks from which pages could not be removed without it becoming obvious. The electronic age has changed the way that laboratory notebooks are kept, but there are still some rules that should be followed.

The purpose of a laboratory notebook is to provide us with all the information we need to reconstruct our experiments for publication and future experimental design. It can also be viewed as a legal document for purposes of studies done in the law-enforcement arena or in cases of scientific fraud. It is important to learn the policies of the laboratory with respect to how data should be recorded and catalogued (e.g. are hand-written accounts required or are electronic notes acceptable with periodic transmission to the lab director?). There are at least two types of approaches to keeping a laboratory notebook: diary or topical. In a diary-style notebook, chronological entries are made daily on what happened including the procedures used and unusual circumstances or observations. In the topical-style notebook, separate entries are made for different topics and/or procedures.

The types of things that we should include in our laboratory notebook can also vary. Pirsig (2008) recommends the following points:

- Statement of the problem
- Hypotheses
- Experiments to be pursued
- Expected results
- Actual results
- Conclusions

This approach is more a diary type and emphasizes that science can be unpredictable and that it is important to carefully observe what happens in the real world compared to what we imagine will happen.

In my laboratory notebooks, I prefer the following items:

- Table of contents
- Detailed description of the methods used
- Results (raw and summary)
- Observations including anything that was unusual
- Conclusions
- Suggestions for further work

See Table 7.2 as an example of the display of raw data for a biochemical time course. This approach is more topical and tends to segregate everything by the method used. As Feynman (1997) suggests above, we must look at our data from a broad perspective. Follow the style preferred by your major professor. If there is no preferred style, select one that best fits your personal style.

Table 7.2 Raw data from a timecourse of the effect of preincubation of flounder muscle microsomes with phospholipase A_2 on enzymic (E) and nonenzymic (N) lipid peroxidation. Note that the labeling is more informal than would appear in a manuscript, but everything must be clear to the investigator at the time the data are being incorporated into the manuscript

Data collected 9–13 2:00–5:00 PM

System	Preinc. (min)	Lipid ox inc. time (min underlined)													
E	0	5	10	15	20	25	30	35	40	45	50	55	60	90	120
		0.7	2.5	7.3	13.5	14.8	20.4	25.5	29.6	31.2	35.7	38.3	44	57	67.6
	30	10	20	30	40	50	60	70	80	90	100	110	120	150	180
		1.2	2.1	2.9	2.8	4.5	4.7	6.2	6.8	8.2	8.7	8.9	9.4	11.3	11.9
N	0	0.25	0.75	1	1.5	2	3	4	5	10	15	20	25	30	
		21.5	29.9	28.9	30.2	30.7	30.6	31.6	32.8	35.6	35.5	36.1	37.7	39.3	
	30	0.25	0.5	1	2.5	5	7.5	10	12.5	15	20	25	30	45	60
		1.8	2	1.8	1.5	2	4.3	17.2	19.1	19	19.1	18.7	19.3	19.3	18.9

One last point: Do not ever put off recording the necessary information. The longer we wait to record what we have done, the harder it will be to reconstruct what really happened. When going back six months after conducting an experiment to write it up, it may become very important to know what day a particular procedure was modified or what day the mass spectrometer was serviced. The more detailed and the more accurate the information is recorded, the more meaningful our data and our analysis of that data will be.

A note on Qualitative Data

The data collected by food scientists tend to be physical (food chemists and food engineers) or biological (food microbiologists and postharvest physiologists) with some social science (sensory scientists and food economists). We are much more comfortable with quantitative data than qualitative data. We tend to believe that the only valid data are numerical data. There are times, however, when qualitative data are more meaningful. Deming (2000), the father of quality, cautioned against an over-reliance on numbers alone indicating that "The most important things cannot be measured."

There are disciplines that are primarily qualitative in nature. Recording, handling, and interpretation of qualitative data are more difficult than quantitative data, but they can provide us with information that cannot be determined with numbers alone. The three primary types of qualitative data consist of interviews, observations, and documents (Patton, 2002). These types of data collection have been developed to observe humans in activities that may be difficult to quantify and are gaining acceptance in the social sciences. Some human-activity systems related to food science include consumer behavior (Dubost et al., 2003), fruit grading (Studman, 1998), and the handling/ distribution system from field to consumer (Florkowski et al., 2009).

Focus-group studies represent a special type of group interview used by some food scientists. Guidelines in conducting focus groups have been developed by Kreuger and Casey (2000). Focus groups are typically broadening exercises to identify a range of perspectives. Well-conducted focus groups will provide a broad section of responses on a product such as critical purchase and consumption attributes for fresh mangoes and peaches in Table 7.3.

For example, a systems approach to better understand handling and distribution of fresh fruits and vegetables (Florkowski et al., 2009) requires use of both quantitative and qualitative data. Tracing changes in quality over time requires quantitative data. Documentation of the system requires extensive collection of qualitative data from all players within the handling system (Prussia and Hubbert, 1991).

The Feynman (1997) admonition particularly applies as well to qualitative data. When collecting these data, we must be very careful to listen to all comments and not "cherry pick" what we wish to hear. We need to report all sentiments or we will get trapped in a hermeneutical circle (Rorty et al., 2008) as discussed in Chap. 4.

Table 7.3 Product attributes critical to purchase and consumption of mango and peach as determined by focus-group interviews (Malundo, 1996)

Product	Critical purchase attributes	Critical consumption attributes
Mango	Color	Flavor
	Size	Mouthfeel
	Firmness	Juiciness
	Aroma	Flesh color
	Fibers	
Peach	Color	Flavor
	Size	Mouthfeel
	Firmness	Juiciness
	Aroma	

The greatest benefit of qualitative research is identifying fresh ideas that provide new insight into a problem. The greatest danger of qualitative research is confirming personal biases of the investigator without a willingness to stretch the mind.

References

Deming WE (2000) Out of the crisis. MIT Press, Cambridge

Dubost NJ, Shewfelt RL, Eitenmiller RE (2003) Consumer acceptability, sensory and instrumental analysis of peanut soy spreads. J Food Qual 26:27–42

Feynman R (1997) Surely you're joking, Mr. Feynman! W.W. Norton, New York

Florkowski WJ, Prussia SE, Shewfelt RL, Brückner B (eds) (2009) Postharvest handling: a systems approach, 2nd edn. Academic, San Diego

Kreuger RA, Casey MA (2000) Focus groups: a practical guide for applied research, 3rd edn. Sage Publications, Thousand Oaks

Malundo TMM (1996) Application of the quality enhancement (QE) approach to mango (*Mangifera indica L.*) flavor research. Ph.D. Dissertation, University of Georgia

Merriam-Webster (2009). Online Dictionary http://www.merriam-webster.com/

Patton MQ (2002) Qualitative research and evaluation methods. Sage Publications, Thousand Oaks

Pirsig R (2008) Zen and the art of motorcycle maintenance: an inquiry into values. William Morrow Paperbacks, New York

Prussia SE, Hubbert C (1991) Soft system methodologies for investigating international postharvest systems. ASAE Tech Paper 91–7050

Rorty R, Williams M, Bromwich D (2008) Philosophy and the mirror of nature: thirtieth anniversary edition. Princeton University Press, Princeton

Studman C (1998) Ergonomics in apple sorting: a pilot study. J Agri Eng Res 70:323–334

Chapter 8
Processing and Analysis

What are your data trying to tell you?

Herb Hultin

Now that the data have been collected, they must be analyzed and processed. Computers are wonderful instruments to process data. We take them for granted now, but some major professors did not have a computer as an undergraduate. It is only since the 1980s that computers have become common fixtures on professors' desks and only since the 1990s that we had the Internet and email. Ask an old-timer about punching cards and floppy disks, and watch them smile.

Computers are great at organizing and sorting data and at making calculations. We use them to conduct statistical analysis, prepare graphics, in simulation modeling, for word processing, and many other ways. We need to familiarize ourselves with some statistical package. SAS is a common program used in statistics classes and many labs. It is a versatile program, although not that user friendly. Find out what is used in the lab and learn how to use it. In addition to understanding the mechanics of performing a statistical procedure, we must also understand the underlying assumptions of the tests we select and the limits of interpretation of the data.

As we become more reliant on computers, there are dangers that we do not always appreciate. If a computer becomes a black box in which all we do is put data in and receive results out, we may draw inappropriate conclusions. One thing many scientists are losing today is the ability to make careful observations. Find a journal article that was written before 1930. It will not be online but somewhere in the library stacks in old musty, yellowing journals. These articles are characterized by detailed observation with very little quantitative data. Also, we need to be careful of computer-generated graphs. They tend to distort data patterns leading to misunderstandings and misinterpretations. Plots where the origin is not 0,0 can visually exaggerate small differences. Broadening or narrowing the X- or Y-axis can provide very different perspectives (see Fig. 8.1). Hand-plotting on old-fashioned graph paper can help us make sense of our data.

R.L. Shewfelt, *Becoming a Food Scientist: To Graduate School and Beyond*,
DOI 10.1007/978-1-4614-3299-9_8, © Springer Science+Business Media New York 2012

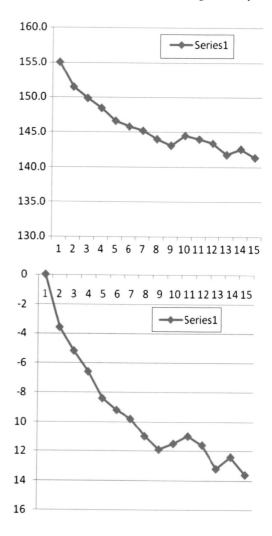

Fig. 8.1 Which of the these plots shows the more dramatic change? See the end of the chapter for the answer

We usually analyze data by statistics and according to a prearranged experimental design as discussed in Chap. 6. Instruction in statistics is beyond the scope of this book, but it is absolutely critical to know and understand the terms in Chap. 6. When using *Analysis of Variance* (*ANOVA*), we will probably use additional techniques to further analyze our data. We should take the summary of our results back to the same statistician consulted in selecting the experimental design. Before going, we should have our data clearly organized and summarized by raw data and means for each of our treatments (see Table 8.1). We must clearly state what was previously negotiated and the resultant constraints as the specifics may have been forgotten, assuming that the full power of the model is being used.

Some of the frequently used statistical techniques used to follow up on *ANOVA* are:

Mean separation (post-hoc comparisons) techniques can be used to determine the treatment effects (e.g., storage conditions, Table 8.2). Some techniques include

Table 8.1 Mean consumer scores (superior = 2, acceptable = 1, unacceptable = 0) for 44 selections of sweet onions and the % of consumers who rated the onions superior, acceptable, and superior plus acceptable

Selection	Mean	% Superior	% Acceptable	% Superior + acceptable
1	1.40	47	47	93
2	1.03	30	47	77
3	1.07	23	60	83
4	1.07	27	53	80
5	1.10	30	50	80
6	1.23	37	50	87
7	1.40	50	40	90
8	1.07	27	53	80
9	1.03	23	57	80
10	1.17	37	43	80
11	1.17	30	57	87
12	1.23	33	57	90
13	1.43	53	37	90
14	1.50	57	37	93
15	1.27	37	53	90
16	1.33	47	40	87
17	1.30	40	50	90
18	1.40	47	47	93
19	1.17	37	43	80
20	1.40	43	50	93
21	1.17	40	37	77
22	1.37	47	43	90
23	1.50	53	43	97
24	1.37	43	50	93
25	1.37	53	30	83
26	1.37	43	50	93
27	1.37	47	43	90
28	1.13	33	47	80
29	1.47	50	47	97
30	1.00	30	40	70
31	1.37	47	43	90
32	1.07	27	53	80
33	1.37	53	30	83
34	1.47	50	47	97
35	1.37	53	30	83
36	1.37	47	43	90
37	1.40	40	60	100
38	1.47	57	33	90
39	1.50	50	50	100
40	1.10	23	63	87
41	1.23	33	57	90
42	1.33	37	60	97
43	1.30	40	50	90
44	1.23	37	50	87

Table 8.2 Effect of ripeness at purchase on texture of early-season peaches when stored at cold storage (5°C, 5 days), room storage (20°C, 5 days), or controlled ripening (5°C, 3 days plus 20°C, 2 days). Sensory texture was evaluated on a 150 mm scale with a higher value corresponding to a firmer peach as perceived in the mouth. Values in the same column followed by the same letter are not significantly different by Duncan's mean separation ($p < 0.05$)

	Cold storage	Room storage	Controlled ripening
Ripe	52b	45a	44a
Partially ripe	62ab	58a	40a
Unripe	86a	51a	39a

Duncan's multiple range test, Tukey's studentized range test, and *LSD (least significant difference)*. Among the various techniques, LSD is more sensitive, detecting more differences among treatments, and Tukey's is more conservative, detecting fewer differences among treatments. Duncan's means separation should only be used for discrete variables such as a cut of meat and not continuous variables such as cooking time or temperature.

Regression techniques are used to develop the mathematical relationships between two or more variables. *Linear regression* plots the linear relationship of one *variable* as function of another *variable* usually on an *x/y* plot. *Multiple linear regression* develops a linear equation of one variable as a function of many variables. *Nonlinear regression* techniques incorporate interaction, quadratic and other terms that predict a single variable. In regression, often multicollinearity (correlation among independent variables) reduces the precision of the parameter estimates. Several approaches can be used to overcome this problem, including *stepwise regression* analysis, *forward regression* analysis, and *backward regression* analysis.

Correlation, one of the most used and abused statistical analyses, relates the mathematical relationship between two or more *variables*. If two *variables* increase concurrently, they are *directly correlated*. When one *variable* increases as the other decreases, they are *inversely correlated*. We tend to become overly impressed with high *correlation coefficients* ($r = 0.90$ and higher), but the importance of a correlation coefficient is its probability, testing the null hypothesis that $r = 0$, which is usually given with the coefficient and usually ignored.

Multivariate techniques such as *cluster analysis* and *principal component analysis* (*PCA*) can be very useful in making sense of huge datasets, particularly in comparing data from distinctly different types of analyses such as sensory testing and chemical analysis. They essentially recognize patterns in the data, can be used to develop relationships, and draw inferences that are difficult to obtain any other way. When misused, however, they result in excellent visuals without providing insight into how to use the information to solve a particular problem.

Precautions

There are many other statistical techniques that can help tease out differences or better understand the complexity of our data, well beyond the brief discussion in this chapter. Some things to be particularly wary about include

- Making sure that the statistical techniques will answer the critical questions posed by our research or meet our experimental objectives—too many methods are adapted from previous research without careful regard for the difference in objectives
- Designing experiments more complex than they need to be such that we are not able to generalize the results or determine the next logical objective of our research
- Using a technique that none of our reviewers will understand (several years ago, reviewers were unlikely to reject what they did not understand, but now the bias appears to reject anything they do not understand—another reason for having a statistician on the advisory committee and as a coauthor on any manuscripts with novel applications of techniques unfamiliar to our reviewers)
- Interpreting a highly significant correlation as implying cause and effect, but such correlations might be the result of the change in one variable causing the change in a second variable, or a third variable (perhaps unmeasured) causing the effect in both of the observed variables, or a mere coincidence

Some frequent mistakes made in using mean separation include

- Analyzing means across one variable and then across the other variable without partitioning the SS (*sum of squares*) which is analogous to spending $50.00 on a meal at a good restaurant and then taking the same $50.00 to buy groceries (most of us can only spend the $50.00 once)
- Separating the means of each treatment within an interaction to compare all means (yes, we can trick our SAS program into performing that function, but that does NOT make it valid!)
- Mislabeling the legend to state that "All means in the same column with a different letter are significantly different from each other ($p < 0.05$)" when the proper terminology should be "All means in the same column with the same letter are not significantly different from each other ($p < 0.05$)"

See Table 8.3 for illustrations of these types of mistakes.

In many fields, modeling is an excellent tool to determine trends and patterns. Models can be built to predict effects, but these models must be validated. Statistical analysis is not the only way to analyze data. Other methods include yes/no answers. Enzyme kinetics and Arrhenius plots are two other ways of generating data without using statistical analysis. Most of what we do in food science, however, has enough biological variability to require statistical analysis.

Interpretation of data should be done in terms of the original objective. Do they support our hypothesis? If so, how do they support it? If not, why not? What are the limitations of our data? What are our data trying to tell us? What is the best way to present these data?

Make sure to consult a good statistics textbook such as Mason et al. (2003), Ott and Longnecker (2008) or the book used at school. Refreshing ourselves on some key points in the book before consulting with a statistician will probably lead to a more productive session.

RULE # 8
Don't sell your important textbooks. They come in handy later!

Table 8.3 Effect of ripeness at purchase on texture of early-season peaches when stored at different temperatures. Sensory juiciness was evaluated on a 150 mm scale with a higher value corresponding to a firmer peach as perceived in the mouth. Values in the same column followed by a different lower-case letter or in the same row or followed by a different upper-case letter are significantly different by Duncan's mean separation ($p < 0.05$)

	5°C	10°C	20°C
Ripe	83 ± 5.2aXY	96 ± 4.2aX	74 ± 5.2bY
Partially ripe	84 ± 3.6aX	79 ± 3.1bX	76 ± 5.5bX
Unripe	53 ± 6.4bY	83 ± 4.4bX	92 ± 4.4aX

CAUTION: There are at least three errors in this table. Can you find them? Answers are at the end of the chapter

Answers to Questions Raised in Fig. 8.1

Both are representations of the same data. The top plot is the change of weight by the author during 15 weeks on the Sonoma Diet (Gutterson, 2011). The bottom plot records weight loss during the same 15 weeks. The bottom plot looks more dramatic because the data are displayed over a narrower range of numbers. When displaying data, we should be sure not to fool our readers or ourselves!

Answers to Questions Raised in Table 8.3

Errors include (1) the mean separation test was performed both down and across—unless the SS are partitioned, this is not a meaningful test, (2) Duncan's multiple range test is not valid for continuous variables like temperature—if the means are to be separated across by temperature, either use mean separation tests like LSD or perform regression analysis, and (3) the phrasing of the mean separation lettering is incorrect as it indicates that the value followed by XY is significantly different than the values followed by X and by Y. Also, we do not know whether the letters a & b or X & Y are designating differences down or across. In addition, some reviewers do not like to see the reporting of both the standard deviation and a means separation. Other reviewers like to see both statistics.

References

Gutterson C (2011) The new Sonoma diet: trimmer waist, more energy in just 10 days. Sterling Publishing, New York

Mason RL, Gunst RF, Hess JL (2003) Statistical design and analysis of experiments, with applications to engineering and science. Wiley, Hoboken

Ott RL, Longnecker MT (2008) An introduction to statistical methods and data analysis. Duxbury Press, Pacific Grove

Chapter 9
Knowledge Dissemination

What makes words so powerful is that they enrich life by expanding the range of individual experience.

Mihaly Csikszentmihalyi (1996)

Writing is thinking on paper.

William Zinsser (2006)

No research is completed until it has been written up. Scientists establish their reputations on the basis of numbers of refereed publications and funded grants. Grant funding is largely based on numbers and quality of refereed publications. This chapter will cover the process of writing a manuscript, submission to a journal, review by referees, the decision of it by the editor, revision and resubmission of accepted manuscripts, and dealing with rejected manuscripts. An accepted manuscript is described as being "In Press" until it is published and becomes an article.

Preparation

There are many ways to go about writing a scientific paper. I will outline what has worked for me. Talk with others who have written manuscripts and had them accepted to get any hints from them on how to proceed. Remember the major professor rule! The first step in writing a manuscript is collecting all of our materials such as the appropriate literature, procedures, and equipment information and our analyzed data. The next step is to establish the boundaries and specifications of the manuscript. Start by writing out the objective, listing the major results and our tentative conclusion.

Before we go much further, we should select the most appropriate journal for our needs. For many of us that publication is the *Journal of Food Science* published by IFT (see Fig. 9.1). The goal is to get as many scientists in our field to read our article as possible. With advanced electronic searching, interested scientists are more likely

R.L. Shewfelt, *Becoming a Food Scientist: To Graduate School and Beyond*,
DOI 10.1007/978-1-4614-3299-9_9, © Springer Science+Business Media New York 2012

Fig. 9.1 Cover of *Journal of Food Science*, April, 2011. Image provided by Institute of Food Technologists

to find relevant articles in obscure sources, but our chances will be enhanced if our article is published in the most appropriate journal. First, we can identify the journals that publish work like the one we plan to author. In what journal are the scientists we are interested in reaching most likely to come across our article? For which journal are we most likely to get a fair review? Journal prestige is also an important consideration. The more prestigious the journal, the more respect our article will receive by other scientists, but the chances our manuscript will be rejected also increases with increasing prestige. Another consideration is turnaround time. Most journals are working to speed up the review process. Some articles will be published within months of submission. Others take years between submission and publication. Before choosing the appropriate journal, check on special requirements. Many journals are published by scientific societies and require that at least that one author is a member of that society in good standing (paid up dues).

Next, select a tentative title for the manuscript. Choosing a title is both difficult and important. Scientists use titles and keywords to determine which articles to read and which to ignore. The title should

- Accurately reflect the research it describes
- Incorporate major keywords/concepts
- Be logical but brief
- Be appropriate for journal it will appear in
- Capture audience interest

Titles beginning with "The effect of ..." like some of my early ones just would not cut it.

Then we need to determine authorship of our manuscript. Authors of a particular article are usually those who have worked on the research from the shoulders up- contributed to definition of the problem, design of experiments, development of methods used beyond what is currently available in the literature, or analysis and interpretation of the data. Generally the list of authors includes the major professor and others intimately involved in the thought process. Persons who collect data are not generally considered worthy of authorship unless they contributed intellectually. The proper order of authorship varies by field, but the first author is usually the person with the greatest involvement and who writes the first draft of the manuscript. The last author is generally the major professor of the first author or the laboratory director responsible for obtaining funding for the project. Other authors are commonly listed in order of contribution to the project. The first author submits the first draft of the manuscript to all authors who make suggestions and return the manuscript for revision. In a multi-authored manuscript, this process may take much iteration. The major professor will probably want to get first crack at the manuscript before it goes to the other coauthors.

RULE # 9
Authors on a manuscript should include anyone who has contributed to it from the shoulders up.

The format of the manuscript is based on the requirements of the journal. Before starting the writing process, access the instructions to authors for the journal (for the *Journal of Food Science*, see Institute of Food Technologists, 2011). Also it is a good idea to read several recent, representative articles from the journal to see how the instructions are applied. Chances are that several of these articles are handy because we have been using them to understand the research area. If there are very few of these articles in our collection, we should reconsider whether this journal is the most appropriate one for our manuscript. The last step is to prepare an outline based on the journal format and the information we collected above.

Writing Process (An Alternative Approach)

Now we are ready to start writing. The process I have found most useful in my writing is counterintuitive, but I have found it to be effective. Use whatever process works best. We can start writing by composing figures (see Fig. 9.2) and tables (see Table 9.1) as we evaluate the data collected. Figures are best when they convey

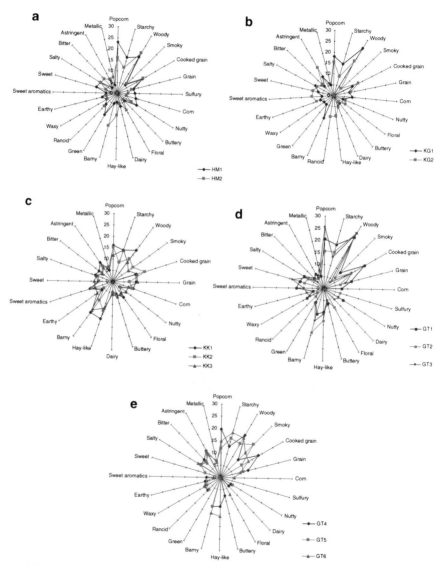

Fig. 9.2 Spider-web plots from Limpawattana et al., 2008. (Spider plot of the mean intensity of descriptors found in (**a**) scented rice, (**b**) glutinous rice, (**c**) black rice, and (**d** and **e**) premium rice. See sample codes in Table 1.). Reprinted with permission from Institute of Food Technologists

Table 9.1 Table of rice cultivars studied in Limpawattana et al., 2008. Reprinted with permission from Institute of Food Technologists

Code	Cultivar	Type	Pedigree/line
HM1	Hyangmibyeo-1	Scented	Suwon 393
HM2	Hyangmibyeo-2	Scented	Suwon 413
KG1	Hwasunchalbyeo	Glutinous	Suwon 384
KG2	Hangangchalbyeo	Glutinous	Milyang 167
KK1	Heukjinjubyeo	Black	Suwon 477
KK2	Heuknambyeo	Black	Suwon 415
KK3	Heukkwangbyeo	Black	Iksan 427
GT1	Ilpumbyeo	Premium	Suwon 355
GT2	Taebongbyeo	Premium	Cheolwon 59
GT3	Hwasangbyeo	Premium	Suwon 330
GT4	Gopumbyeo	Premium	Suwon 479
GT5	Samkwangbyeo	Premium	Suwon 474
GT6	Choochungbyeo	Premium	Akkibari

a clear picture of the interrelationships of the data. Tables function better to convey large datasets that defy graphic presentation or would take numerous figures to present the same information. A figure or table must be able to stand on its own without referring directly to the text. Thus, the figure or table legend must be clear and concise (see Fig. 9.2). For ideas at the detail required, look at the representative articles from the selected journal. Photographs are useful when a brief explanation cannot convey the information, but they should not be overused. Some journals permit color photos; others do not. Consult the instructions for authors if color photos are important. At this point, count the number of figures and tables planned to determine if the numbers fit into the typical range for the journal. If this number is less than typical for the journal, expand the dataset, run some additional experiments, or select a journal that publishes shorter articles. If this number is more than typical for the journal, consider partitioning the dataset into two manuscripts, eliminating some of the data that do not directly relate to the objective written earlier, or finding an appropriate journal that publishes more in-depth articles.

The next section to write is the *Results and Discussion* as it is constructed around the figures and tables. The first decision to make is whether to write them separate or together. Some journals make a specific requirement. Others permit the author to choose. Again, consult representative articles to see if there is a preferred format even if there are no specific guidelines. Generally writing separate *Results* and *Discussion* sections are best when each figure or table provides a specific piece of information but the figures do not build on each other. In this case, present each figure or table in the *Results* section highlighting the key points without drawing any inferences. Then in the *Discussion* section, draw inferences from the data presented referring directly to the appropriate figures and table. A combined *Results and Discussion* section works best when each figure or table builds on the previous one. In this approach present the results of the figure or table and draw inferences. Then introduce the next figure or table, draw the inferences and relate them to the previous inferences. Go back and read some separate and combined *R&D* sections to get a feeling for the two different approaches. One of the differences between *Results* and *Discussion* is that *Results* focus on the data presented in the manuscript and *Discussion* places the data and inferences in context of the scientific literature. Thus, cite previous research in *Discussion* but not in *Results*.

Then it is time to write the *Summary and Conclusions*. A summary is a brief statement of main points. It is not the time to rewrite the whole manuscript or to add new information to the *Results and Discussion*. The conclusions provide our interpretation of the meaning and importance of the study. Like the summary, the conclusions should be brief. *Conclusions* should not repeat the material in the *Discussion* section, but it should distill down our discussion to the manuscript's essence.

At this point it is time to write the *Materials and Methods*. Only the methods actually used to produce the reported results should be described here. Many beginning authors make the mistake of describing all the methods used in the study whether the results were reported or not. Describe clearly, but concisely, any materials and special handling used for these methods. Also mention any specific instruments or equipment used complete with manufacturer name, model, and city of the manufacturer. In the description of the procedures, keep it short. If following a previously published procedure, cite the publication and do not repeat the description. Clearly and concisely state any modifications and the reasons for them. The *Materials and Methods* section is the one most editors recommend for cutting to reduce manuscript length.

The next section is the *Introduction*, sometimes referred to as the *Review of the Literature*. In this section, focus on the journal articles that provide the basic background for the results reported. It is important to note that a journal article is not a report on how the problem was approached. Rather it is a carefully constructed argument on what was learned by conducting well-designed experiments. Each section should reenforce the other sections. In the *Introduction*, set up the reader for the main message of the article and cite pertinent review articles that cover the area studied. It is not necessary to cite individual studies cited in these review articles unless they apply directly to what was done. A research article that presents original data is not one that extensively reviews the literature. Also cite pertinent, specific research studies that provide a direct background to the study. The last paragraph is generally a statement of the objective (what the British call aim) of the study which may include a long-term goal of this line of research in the senior investigator's laboratory. The objective must be clearly stated. The objective is related directly to what was found; it is not necessarily the objective identified when beginning the experiments. The next time when reading a journal article, look for the objective in the last paragraph of the *Introduction*.

The last section written is the first one that the reader sees, the *Abstract*. *Abstracts* are short, frequently less than 250 words (see Fig. 9.3 for an example). The shortest section is usually the most difficult to write. Many abstracts begin with a background sentence, but some omit it. Every abstract should contain the objective (as stated in the *Introduction*), the primary treatments, a brief statement of major results, and the primary conclusion. The *Abstract* may be the most important section of the

Flavor is a key factor contributing to consumer acceptance and repeat purchase of rice. Plant breeders focus on production yield and ignoring quality traits because there are no readily useable tools to evaluate quality. A systematic approach is needed for rice breeders to select rice with favorable flavor traits. Descriptive sensory analysis combined with chemical analysis provided an insight of sensory significance to interpret chemical data for a better understanding approach of rice flavor. This study was aimed to develop prediction models for sensory descriptors based on the volatile components derived from the gas chromatography–olfactometry (GC–O) that would be useful to help select rice cultivars containing a satisfactory flavor to produce improved quality in rice breeding programs. Thirteen Korean specialty rice samples were evaluated for their flavor components using descriptive analysis and GC–O. Nineteen aroma attributes in cooked specialty rice samples were evaluated by 8 trained panelists and statistically correlated to the concentration of aroma-active compounds derived from GC–O analysis. Prediction models were developed for most aroma descriptors including popcorn, cooked grain, starchy, woody, smoky, grain, corn, hay-like, barny, rancid, waxy, earthy, and sweet aroma using stepwise multiple linear regression. (E,E)-2, 4-decadienal, naphthalene, guaiacol, (E)-2-hexenal, 2-acetyl-1-pyrroline, 2-heptanone contributed most to these sensory attributes. These models help provide a quantitative link between sensory characteristics of commercial rice samples and aroma volatile components desirable in developing a rapid analytical method for use by rice breeders to screen progeny for superior flavor quality.

Fig. 9.3 *Abstract* from Limpawattana et al., 2008. Reprinted with permission from Institute of Food Technologists

manuscript as that will be the only part of a published article that many readers will see. *Abstracts* appear in other printed and online sources that do not include the rest of the article. Make sure that it includes all the keywords that the readers are likely to key in. The keywords lead the searcher to the *Abstract* and the *Abstract* frequently determines whether a potential reader will look at any of the rest of the article.

Finishing touches include the *Acknowledgments* of technical assistance (who else helped prepare samples or conduct the research—usually those who contributed their labor but not their intellect), statistical analysis (be sure to check with the statistician with the final copy as that person's reputation is at stake if the statistical analysis has been done incorrectly), materials (someone who provided special cultures or experimental samples, etc.), and, most importantly, the source(s) of funding for the study. Also, complete the list of *References*. Only articles that were cited in the text of the manuscript (do not forget any articles cited in the figures and tables) should be included. Make sure they are in the right format prescribed by the journal. This section is the one most prone to mistakes so check and recheck it. A quick analysis of the references would also be a good idea. No references cited from the journal we plan to submit suggests that we may be selecting the wrong journal, particularly, if the manuscript cites many references from another single journal. If almost all of the articles cited are more than five years old, we may not have done a thorough job reading the recent literature. If almost all of the references are less than five years old, it may not have the proper historical perspective.

Some writing hints include:

- Avoid jargon but be specific
- Allow enough time to write as writing always takes longer than expected
- Find the proper environment to write that is free from distractions
- Make sure that nouns and verbs agree and that the verb tense is consistent
- Remember that many readers may not speak the language of the article so avoid complex words
- Try to communicate the key message to the proper audience
- Use clear, concise, and scientific language

Those who are still insecure should find a good book on writing that fits their needs and style such as those by Day and Gastel (2006), Zinsser (2006), Matthews and Matthews (2007), Katz (2009), or Wallwork (2011). The Wallwork book is particularly useful for those students whose primary language is not English. Rather than trying to read all of these and other books on writing, scan several and choose one to serve as a guide. Buy a copy and keep it handy.

The Review Process

Every manuscript is reviewed and revised many times before it is ever published. Anyone sensitive to somebody messing with their copy needs to get over it. Realize that the manuscript process works at a very slow pace. Be the first reviewer and reviser of any manuscript written. After completing the first draft, rest for a time (a few hours to a few days) before reading it again. Review it for spelling, grammar, style, and content. Make sure that the objective is consistent with the results and the results are consistent with the conclusions. Is the title still appropriate for the manuscript? Does everything fit together?

The next reviewers and revisers are the coauthors. Forward them a clean copy and ask for comments. If acknowledging a statistician, this is the time to forward it for comments. It is generally advisable to give any reviewers a desired completion date, preferably at least a week. Many departments and programs have an internal review of manuscripts that will bear that department's name. Become familiar with the rules and allow time for adequate internal review.

Once it has met all the internal requirements, it is time to prepare the journal submission. Read and follow the guidelines for submission very carefully. Does it require hard copies or is it an online submission? What information is needed in our cover letter or submission message? The manuscript will go to the editor who will select the reviewers, usually two or three experts in the field. Editors may limit reviewer selection to the editorial board (usually published on the website and on the cover of each issue) or may use a wide range of experts who frequently publish in the journal, are cited in the references or are colleagues of the editor who are familiar with this topic. The editor receives the comments and recommendations of the reviewers and makes a decision as to whether to accept, accept with revisions, or reject a manuscript. If there is a split recommendation, the editor may send the manuscript to a referee for another opinion. Once the decision is made, the editor will send a decision with reviewer comments.

If the manuscript has been rejected, carefully review all of the comments provided. At this point we must make a decision on (1) completing a major revision and resubmission to the same or a more appropriate journal, (2) conducting further experiments to answer the critical questions followed by a rewrite and resubmission, (3) filing the manuscript away for a later date, or (4) throwing it away and using the experience as a learning opportunity.

If the manuscript is accepted (with or without revisions), it is time to celebrate. Few events in a scientist's professional life are as exhilarating as notification of that first accepted manuscript. If revisions are required, look over the recommendations very carefully. It may appear that one reviewer has completely missed the point and just does not understand. In this case, it may be that the reviewer is not well qualified in this specific area, but it is more likely that our wording is not clear. If the comments of a reviewer are not clear, consult coauthors to see how best to revise the manuscript. Make all of the revisions that are reasonable or can be lived with. If there are items that are questionable, clearly state the reasons for not making the revision(s). An editor is usually willing to give an author the benefit of the doubt on one or two issues as long as the author is willing to make adequate revisions in other areas. Before resubmission, make sure that all authors are comfortable with the changes made and the response to the editor.

An accepted manuscript will be acknowledged by a letter of acceptance and the manuscript will go to typesetting. A final proof of the manuscript is sent with a short turnaround time to determine if there are any errors. Drop everything and carefully screen the proof to make sure there are no errors. Do not forget titles, headings, and tables. When I was in graduate school, there was a spelling error in the heading on each odd-numbered page of one of my colleagues. We were merciless in our teasing. It turned a triumph into an embarrassment. Checking the proof is the time to catch errors not to change content. Journals sometimes charge for large changes in the manuscript at proofing times. Also, at this time there will be a bill for publication charges and reprints. Present the bill to the person who runs the lab (major professor?) to see how this bill is to be paid.

Writing manuscripts may be the most significant task for a scientist as nothing contributes to reputation among peers as the scope and quality of published articles.

References

Csikszentmihalyi M (1996) Creativity: flow and the psychology of discovery and invention. Harper Perennial, New York

Day RA, Gastel B (2006) How to write and publish a scientific paper. Greenwood Press, Westport

Institute of Food Technologists (2011) Author style guide for IFT Scientific journals. http://www.ift.org/Knowledge-Center/Read-IFT-Publications/Journal-of-Food-Science/Authors-Corner/Author-Guidelines.aspx

Katz MJ (2009) From research to manuscript: a guide to scientific writing, 2nd edn. Springer, Dordrecht

Limpawattana M, Yang DS, Kays SJ, Shewfelt RL (2008) Relating sensory descriptors to volatile components in flavor of specialty rice types. J Food Sci 73:S456–S461

Matthews JR, Matthews RW (2007) Successful scientific writing: a step-by-step guide for the biological and medical sciences. Cambridge University Press, New York

Wallwork A (2011) English for writing research papers. Springer, Dordrecht

Zinsser WK (2006) On writing well: the classic guide to writing nonfiction, 30th edn. HarperCollins, New York

Part II
Maturation of a Scientist

Experience is the best teacher, and the rougher the experience the deeper the learning.

Richard Sands

Chapter 10
The Scientific Meeting

It's not what you say. It's what they hear.

Red Auerbach

Our hotel is so primitive it doesn't even have a connection to the Internet.

student talking to her professor as overheard at the 2009 IFT Annual Meeting in Anaheim

There are many types of scientific meetings. The most familiar meetings are those sponsored and convened by a scientific organization. These organizations can be international, national, regional, or state. The primary organization for food science is IFT, Institute of Food Technologists. Other important organizations for food scientists are ASM (the American Society for Microbiology) and ACS (the American Chemical Society). A list of organizations of interest to food scientists can be found in Table 10.1. IFT is a unique organization in that it combines a scientific meeting with a trade show (see Fig. 10.1). Every food scientist needs to attend the IFT meeting at least once in a lifetime. It must be experienced to be appreciated. Every practicing scientist should become a member of a scientific organization. University food scientists usually belong to at least one additional organization that is more aligned with their disciplinary perspective such as ACS or ASM. For postharvest physiologists, ASHS (American Society for Horticultural Science) and ASPB (American Society of Plant Biologists) offer alternatives. Few alternatives to IFT have a major trade show at their meetings analogous to IFT. Industrial food scientists usually divide up responsibilities among relevant societies within the company to cover each of the important areas.

Another type of scientific meeting is organized by a narrowly defined topic. These meetings tend to be much smaller with scientists actively working in the area of emphasis. Such meetings allow an in-depth exploration of the topic without requiring the formation of a society. They can be called symposia, colloquia, conferences etc. Frequently these meetings are international, but they can encompass a much smaller scope. Sometimes they are organized within a society. They provide

R.L. Shewfelt, *Becoming a Food Scientist: To Graduate School and Beyond,*
DOI 10.1007/978-1-4614-3299-9_10, © Springer Science+Business Media New York 2012

Table 10.1 Food-related societies that host scientific meetings and their headquarters location

Acronym	Society Name	Headquarters
AACC	American Association of Cereal Chemists	St. Paul, Minnesota
ACS	American Chemical Society	Washington DC
AMSA	American Meat Science Association	Champaign, Illinois
AOCS	American Oil Chemist's Society	Urbana, Illinois
ASBC	American Society of Brewing Chemists	St. Paul, Minnesota
ASHS	American Society for Horticultural Science	Alexandria, Virginia
ASM	American Society for Microbiology	Washington DC
CIFST	Canadian Institute of Food Science & Technology	Toronto, Ontario
CIFST	Chinese Institute of Food Science & Technology	Beijing, China
IFST	Institute of Food Science & Technology	London, UK
IFT	Institute of Food Technologists	Chicago, Illinois
IUFoST	International Union of Food Science & Technology	Oakville, Ontario
PSA	Poultry Science Association	Savoy, Illinois
RCA	Research Chefs Association	Atlanta, Georgia

Fig. 10.1 Logo for IFT Annual meeting in Las Vegas, June 25–28, 2012. Image provided by Institute of Food Technologists

an excellent opportunity to provide a much more focused view of a specific research topic getting beyond the more general reviews at a meeting like IFT.

Gordon Research Conferences represent another approach to this type of meeting. They organize around a topical area for 100 of the top practicing scientists in the world for a week of technical sessions and social interaction. Gordon Conferences with food applications include those on

- Carbohydrates
- Carotenoids
- Cellular and Molecular Fungal Biology
- Enzymes, Coenzymes & Metabolic Pathways
- Floral & Vegetative Volatiles
- Marine Microbes
- Microbial Stress Response
- Microbial Toxicity & Pathogenesis
- Molecular & Cellular Biology of Lipids
- Muscle: Excitation/ Contraction Coupling
- Oxygen Radicals
- Postharvest Physiology
- Proteins; and many more

For more information, see http://www.grc.org/. Because space is limited at each Gordon Conference, everyone must apply with their credentials to be accepted. They do have a special program for graduate students and postdocs for the conferences and the Gordon Research Seminars which can be accessed at http://www.grc.org/grs.aspx.

Other meetings include short courses (e.g., HACCP held at many universities) and workshops. Short courses involve transmission of knowledge from experts in the field to people who are new in the area or those wanting to upgrade their expertise. Workshops can be similar in nature to short courses or they could be focused on solving a specific issue or problem by experts in the field representing different perspectives. Although the traditional way of convening these meetings is face-to-face, video conferencing and webcasting are now being used for knowledge delivery.

Meeting Activities

Many students think of scientific meetings as a place to party and to listen to technical presentations, but there is much more to a scientific meeting. Meetings include exhibitions, employment interviews, tours, social events, business meetings, and many other activities.

Technical sessions usually consist of oral or poster presentations but can extend to demonstrations, panel discussions, and other formats. The traditional form of technical presentation is the oral version of the scientific paper in which a scientist is allotted 15–30 min to present research results to an audience. Usually 3–5 min is reserved at the end of each oral presentation for questions from the audience. Session moderators have a responsibility to keep these sessions on time, particularly when there are concurrent sessions, so that members of the audience who want to pick and choose papers from different sessions are able to attend the papers of their choice. Although oral presentations were given the most prestige in the 1980s and before, poster sessions are now more popular. The oral presentation is probably best for audiences of 50 or more, but poster presentations are optimal for audiences of 10–50. There will be specific guidelines provided for the size and scope of posters permitted. At IFT, the posters are to be posted for 2 h, and one of the authors is required to be present for at least one-and-a-half h. At other organizations, the posters may be up for the entire meeting with the presenters given a time (such as 2 or 3 h) when they are required to be present for discussions with interested participants. If a poster attracts less than ten interested participants, the presenters probably chose the wrong meeting to present these data.

Not all oral presentations report original data. Some provide overviews on a specific topic and are called symposia, colloquia, etc. These sessions are excellent for learning about a new field, for keeping up with a secondary area of interest, or to see and hear a scientist whose work you have read. Scientists who are keeping up with an area in the literature are frequently disappointed with these presentations at annual meetings of a society as they tend to cover what is already known to practitioners. At the smaller conferences, however, there are few or no concurrent sessions. The papers are more focused, and presenters assume the audience is familiar with the field. Demonstrations of scientific techniques or capabilities of apparatus at scientific meetings can also be useful in keeping up to date in a field. For example, the culinology demonstrations at IFT show how food science meshes with culinary arts.

Exhibitions are usually directed at commercial activity. In a trade show, suppliers interact with manufacturers. At IFT, there are numerous categories of display areas including food manufacturers, packagers, ingredient suppliers, processing equipment suppliers, laboratory instrument companies, consultants, university research laboratories, and publishers. The show affords an opportunity to sample many new products and meet people from food and aligned industries. At meetings without a major trade show, there are usually exhibits by manufacturers of laboratory instruments and publishers of relevant books and journals.

Annual meetings of a society are excellent places to search for a job. Employers are attracted to annual meetings because of the large number of potential employees present and the likelihood that an attendee is more interested in the field than those who do not attend. Although interviewers are searching for people to fill specific jobs, annual meetings can also serve as scouting missions. Most of these meetings have an official employment bureau that sets up interviews between employers and job seekers. Rules for the employment bureau are available in the meeting announcements. Some bureaus permit on-site registration and posting of résumés, but others require pre-registration. Interviews at meetings help an employer screen a large number of potential employees in a short time. Few job offers are made at an annual meeting, but meeting interviews lead to site visits where more extensive interviews are conducted. Job searches are not confined to the employment bureau. There are many other opportunities for employers to view potential employees (e.g., at oral or poster presentations) or for job seekers to study potential employers (e.g., exhibits or social events).

When at the peak of a career, the location of a scientific meeting may make little difference with the exception of how long it takes to get to it and how long it takes to get home. Meeting activities tend to expand to take over the entire time of the meeting. At the larger meetings professional tours are offered to see food-processing facilities or research labs in the area. Sightseeing tours for family members and meeting participants are generally available to get a flavor of local culture. Many attendees take advantage of sampling fare at local restaurants which are usually more enticing than foods at the conference center or hotel. A meeting in a foreign country affords the opportunity to combine the event with a vacation. If the university or company is paying for expenses, however, it is a good idea to make sure that vacation plans don't jeopardize funding. Some universities or research laboratories will allow no travel outside the meeting itself.

Social events are important parts of scientific meetings, but don't confuse them with party time. The primary purpose of the social events is to provide networking opportunities. Nothing at an annual meeting is off the record. These events bring together people with like interests (e.g., alumni, faculty, and students of a specific university) in a more informal setting. The events provide potential employers, colleagues, and collaborators the opportunity to determine if these people are those they would choose to hire or work with. It provides a younger researcher the opportunity to ask questions about ongoing research and maybe learn about advances or difficulties not discussed in the formal presentations. Attendance is required at all meals at Gordon Research Conferences to foster informal communication. More scientific information is exchanged in small groups across the tables during the meals than in the oral or poster sessions. The formal sessions are held in the morning and the evening with the afternoons left open for specialized group meetings, rafting trips, etc. that foster direct interaction between scientists.

Most societies are confederations of smaller groups called divisions, working groups, etc. Such groups have officers and business meetings at the annual meeting. They are frequently organizers of technical sessions in their particular area of interest. They provide opportunities for networking and leadership. They also help members keep up with other conferences, employment opportunities, and grant programs in a specific area.

Meeting Planning

I recommend some strategic planning before going to a scientific meeting. Students who have scientific results to share should apply for a presentation. Consult the guidelines for a call for papers (IFT, 2011). Note the program tracks at the IFT meeting listed in Table 10.2. All abstracts are evaluated in the context of these tracks. Acceptance requires submission of an abstract and other important information, usually to be completed online. It may require membership. Make sure to consult with all potential coauthors before submitting an abstract. Then wait to hear if the abstract has been accepted for presentation.

Meetings are expensive: registration, travel expenses, hotel, meals, cab fares, entertainment, and other incidental expenses. Students may have to pay their own way. Check with colleagues who have gone to that meeting before to learn the appropriate dress. For example, blue jeans are NOT appropriate for IFT, and a business suit is NOT appropriate for ASHS. There may be ways to get funds to travel to a meeting. Talk to the major professor, particularly if presenting a paper on that work. The graduate schools at many universities have some limited funds to support travel for graduate students presenting their research at a national meeting. Some societies provide travel grants or provide free rooms to students willing to work in various capacities. Some companies pay students to work their exhibit booths. Food Science Clubs may help fund student expenses of those active club members who participated in the fund-raising activities over the year. Finding budget airfares, traveling in vans with other students, sharing rooms at budget hotels away from the conference location (BEWARE though of exorbitant fees for parking), finding low-cost food or feasting off exhibitor samples, and sharing a cab or limo from the airport with others are just some of the ways to save money. Most meetings have lower pre-registration fees for those who make their commitment early. Study all the rules and regulations before going.

Table 10.2 Submitted abstracts to the IFT Annual Meeting are evaluated within these program tracks. Relevance to a track in each category enhances acceptance of the abstract. These tracks are likely to change from year to year

Core Science Program Tracks	Key Focus Area Program Tracks
Food Chemistry	Education and Professional Development
Food Engineering	Food, Health, & Nutrition
Food Microbiology	Food Processing & Packaging
Sensory Science	Food Safety & Defense
	Product Development & Ingredient Innovations
	Public Policy, Food Laws, & Regulations
	Sustainability

RULE # 10
When traveling take twice as much money and half as many clothes as you think you will need.

To get the full benefit of a meeting, carefully study the technical program. Map out the technical presentations and sessions to attend. Meetings with many concurrent sessions require difficult choices. Carefully study the map of the meeting rooms to have adequate time to move from one session to another. Scientific presentations are important. Attend several oral and poster presentations that are related to specific research interests, but don't make the mistake of missing other important aspects of the meeting just to attend the sessions.

If seeking employment at the meeting, make sure to follow the posted guidelines. Have someone like a professor, a career counselor, or a recent graduate working in the same field review the résumé before submitting it. Carefully plot out a strategy. Identify and prioritize the important aspects of the job such as the type of work, potential for advancement, location, minimum salary, etc. Identify the characteristics that are strengths and the types of characteristics of the interviewing organization. The interview should be a two-way conversation to make sure that there is a good fit. Don't expect an offer to come out of the meeting. The goal is usually an invitation for an on-site interview at the company or institution. Make sure to get contact information from all interviews that were promising. Follow up on all interviews for jobs to seriously consider.

Perhaps the most important aspect of any scientific meeting is networking. Identify at least ten persons to meet and the venues most likely to meet them. Some places that are good for meeting people are

- After the oral session where they have presented a paper (never bother a speaker right before the presentation)
- At a poster session presentation
- At a social event of mutual interest
- At a business meeting open to all attendees

Scientific Presentations

If the abstract is accepted to give a paper at a meeting, there are several things to consider. Check with the guidelines to learn how much time is allotted. Make sure to allow a few minutes for questions. Usually these presentations provide an overview of research that has been conducted or a state-of-the-art of a certain technique or process. Check with the session moderator ahead of time to determine if there are any expectations in addition to the posted guidelines. Make sure to tailor the presentation to the audience. A presentation to an IFT audience would be different from one to ACS or ASM.

Oral technical papers are of a more limited duration. IFT sessions usually provide 12 min for the presentation and 3 min for questions. Practice the presentation to ensure that it takes between 11.5 and 12.5 min. Any deviation from the time is rude and unfair to the next presenter. Presentations should be results oriented unless the major objective of the paper is to develop a method. Start with a brief description of the project and the research objective. In general, no more than one or two slides should explain the methodology. At least half of the presentation should focus on the results. Not all of the results need to be presented, just those that relate directly to the main points. Don't necessarily use the whole figures and tables in the written manuscript. If focusing on only a small part of a figure or a table, present only that part. Every paper should have a summary slide and a conclusions slide. Some specific guidelines for an oral presentation include

- Focus on key points, don't get bogged down in details
- Use the slides as note-cards, but have some more detailed notes to refer to if lost
- Practice enough until comfortable with the material but not to the point that it becomes dry
- Deliver the paper in a conversational style; don't read it or present it from memory
- Be prepared for questions, but don't be afraid to admit to not knowing the answer ("That is an interesting point that I will need to think about" is an acceptable answer)

In designing a PowerPoint presentation, start with a slide with the title and the authors of the paper. It is usually not very cool to restate the title of the presentation, particularly if the moderator has already read it. All pictures should be clear and easy to read. Give credit to any source used. Don't use a picture, figure, or table scanned in from another source without permission to use it from the copyright owner. All of the figures and tables should be readable from a distance. Keep word slides simple. Make sure to follow all of the guidelines for online submission of an oral presentation.

Poster presentations should carefully follow guidelines for size of the poster and font. At a recent IFT meeting, a presenter had to leave half the presentation in the carrying container because it did not conform to guidelines. Preparation should be geared to results, have a pleasing overall design, and provide an eye-catcher to draw interest. In general, the format is the abstract, a brief introduction, an outline of the

We were reviewing resumes for a faculty position. One person asked if anyone knew one of the applicants who had very impressive credentials. No one on the search committee had heard of him. The person who posed the question indicated that she had been in charge of the audiovisuals at the IFT meeting, and this applicant had pitched a royal fit because of some glitch with the presentation. She was not favorably impressed with the applicant's attitude, and he was dropped from further consideration for the position.

Fig. 10.2 A precautionary tale

methods, figures and tables of the results, a brief discussion, and a conclusion. In labeling the figures and tables, use short titles that relate directly to the main point(s). They do not need the in-depth figure legends found in manuscripts. Print up several miniature paper copies of the presentation to hand out to interested attendees. On the day of the presentation, arrive early, be prepared with materials to hang it (IFT provides Velcro stickers to fasten it to a fabric background), remain with poster for the prescribed period, and be ready for questions and answers. Poster sessions can become social events, but make sure that socializing doesn't interfere with scientific interchange. Remember, the reason for being there is to present data and to learn from those who have similar interests. Technical visitors might include a potential employer, future mentor, research collaborator, colleague, reviewer of a manuscript or grant proposal, or competitor. Treat each visitor with respect (see Fig. 10.2).

Reference

IFT. 2011. Call for Abstracts. Submission Guidelines. https://www.am-fe.ift.org/pdfs/Tech%20 Research%20Paper%20guidelines12_FINAL.pdf

Chapter 11
Critical Thinking

As the complexity of the world seems to increase at an accelerating rate, there is a greater tendency to become passive absorbers of information, uncritically accepting what is seen and heard.

Neil Browne and Stuart Keeley

As discussed in Chaps. 4 and 9, we must carefully evaluate the articles we read in the scientific literature. Effective analysis of the scientific literature requires special skills, generally grouped as critical thinking. Many sources describe critical thinking (Fisher, 2001; Paul and Elder, 2002; McInerny, 2005; Browne and Keeley, 2011; Burton, 2008). Critical thinking has been defined as

- "...an awareness of a set of interrelated critical questions, plus the ability and willingness to ask and answer them at an appropriate time." (Browne and Keeley, 2011)
- "...the ability to interpret, analyze and evaluate ideas and arguments." (Fisher, 2001)
- "...active, persistent and careful consideration of a belief or supposed form of knowledge in the light of the grounds which support it and the further conclusions to which it tends." (Dewey, 1909 as quoted by from Fisher, 2001)

Immersion in critical thinking can be one of the most liberating or most devastating experiences in a student's career. True critical thinking reveals the uncertain nature of knowledge. Most students become scientists to gain certainty about their world and can become disillusioned when they learn that we don't always really know what we think we know. Nevertheless, science moves forward based on premises, evidence and confirmation.

R.L. Shewfelt, *Becoming a Food Scientist: To Graduate School and Beyond,*
DOI 10.1007/978-1-4614-3299-9_11, © Springer Science+Business Media New York 2012

Important Terms

Before delving into an understanding of critical thinking, we need to define some key terms. We can critically evaluate any information we receive in written, audio or video form, but we are looking for the message behind the form as composed by the author, speaker or designer.

Argument—an interweaving of the evidence and the conclusion
Assumption—a premise, either stated or unstated, that is assumed to be true without verifiable information
Conclusion—the main message being delivered
Evidence—verifiable information presented leading to the conclusion
Inference—logical deductions based on accepting the assumptions and the evidence
Value—deeply held belief as to how things are or how they should be

These definitions are adapted from Browne and Keeley (2011). Since some terms are used to define the other terms, we could become victims of circular reasoning. Critical thinking is difficult to explain, but the concepts usually become clear upon reflection. This discussion may be boring, but if we are going to become effective scientists we must come to terms with critical thinking.

> *I have little patience with scientists who take a board of wood, look for its thinnest part, and drill a great number of holes where drilling is easy.*

> Albert Einstein

Critical Thinking Processes

There are several questions we should be asking ourselves as we evaluate a message critically. Some of these include:

- What is the author or speaker trying to say?
- What level of expertise does the author have?
- Is the message based on data or based on opinion?
- Is the argument testable or verifiable?
- What are the stated and unstated assumptions?
- Are these assumptions valid?
- Has relevant information been withheld?
- Does the argument support or refute previous reading?
- Are the conclusions supported by the evidence presented?
- Are there alternate explanations for the observations?
- What are the values expressed?
- Does the message provide us with new insight into the topic?

Many people think that critical thinking is about shooting down other people's ideas, but that is a gross oversimplification. Critical thinking is mainly about taking someone else's thoughts and weighing them against our own. True critical thinking involves critically evaluating our own ideas and values as much as those of the message evaluated. It is not enough to disagree strongly with the author of these ideas. It is then important to determine the reason for the disagreement and evaluate dispassionately which ideas are superior. Systematic models have been developed to help apply critical thinking to specific applications (Afamasaga-Fuata'i, 2008).

We can classify thinkers in three broad categories. A palm tree bends in the wind to prevailing thought and serves as an avatar for someone who consistently concludes that the message is correct and their own thinking is flawed. Palm trees are impressionable and likely to express the opinions of the last article they read or the last person they listened to. A boulder takes great force to move and serves as an avatar for someone who is never swayed by an argument and convinced that they are always right. A fox is neither swept away by the latest trend nor rigidly dogmatic, but can carefully evaluate each situation learning to adapt and survive.

Critical thinkers develop thought styles (see Grinnell, 1992 and Chap. 1 for a more detailed discussion of thought styles and collectives) that reflect their environment and exposure to ideas. They are not afraid of challenging themselves with new ideas or challenging others with their ideas. They are willing to admit that they are not always right. They read widely and listen carefully. They are in touch with their own values and develop a thought style that reflects these values. They use the thought style to evaluate what they read, see, and hear, but are not afraid to modify their thought style and even reevaluate their values based on new information.

> By a timeless irony, religion, which speaks of brotherhood, has divided men; whereas trade, the vehicle of his self-seeking, has united him.
>
> Colin Thubron (1978)

Cultivating Critical-Thinking Skills

Critical thinking is not something we typically engage in until trained. It involves getting out of our comfort zone. Taking a Philosophy or Education course would be a good idea. Some courses in Food Science departments may feature discussion of current literature in lieu of lectures and a textbook. These courses tend to encourage critical thinking. Discussion seminars, journal clubs and learning communities are also frequently excellent ways to hone critical thinking skills. Seek out those that have a wide diversity of opinion, not a collection of group thinkers. Debating clubs and other organizations that stress looking at issues more deeply can also be beneficial. If none of these options are available, organize a journal club with classmates. If possible, find a professor that is open to new ideas to provide some direction and may serve as a moderator. Advocacy groups and political organizations tend to have a specific agenda and tend to be closed to critical thinking. Their objectives tend to be to find support for their preconceived ideas and flaws in the thinking of their opponents. Opposition research is not critical thinking.

Once your mind is inhabited with a certain view of the world, you will tend to only consider instances proving you right. Paradoxically, the more information you have, the more justified you will feel in your views.

Nassim Taleb (2010)

One way to train ourselves on critical thinking is to get a good book on the topic, such as one listed in the chapter references. Another is to start employing a critical-thinking regimen to some of our reading, watching, and listening. It might or might not relate to our area of research. Write down the questions above on a piece of paper or enter them on the computer screen. Evaluate a short article, you-tube video or newscast. Concisely state the argument, assumptions, conclusion, evidence, inferences, and values in that short piece. Provide answers to each of the questions. We should describe how it has shaped our thought and then set the piece aside for a day or two. Has the meaning of it changed any since we first looked at it? Concept maps like the one developed by West et al. (2000) can help us develop our ideas.

Apply this approach to progressively longer articles, both professional and non-professional. Use it on familiar topics. It is easy to have opinions without knowl-edge. We can test ourselves with familiar topics and be just as critical of our own thoughts as those of others. We should find someone else who is interested in honing their critical thinking skills and a rigorous exchange of ideas. These methods will be tedious at first, but with more practice, we will find ourselves asking and answering the important questions without needing to write them down.

Critical Thinking for the Food Scientist

When starting to read articles for knowledge in the field, we need to evaluate each article critically. Reading without note-taking is nonproductive. Development of a knowledge base in our research area is essential before full-blown critical thinking. Our notes at first may be sketchy and very broad as we learn what is generally accepted and uncover unresolved issues. Separating out data-based conclusions and speculation is a trick we will need to master. As we read, we will begin to see simi-larities in the style and perspective of each lab studying the issue. Points which are generally accepted by all research groups are the basis for the thought collective. Differences between groups represent the clash of thought styles or interpretation. Differences with direct relevance to our research require careful reading of the specific articles these authors cite and a strong basis for the design of meaningful experiments that answer ambiguities.

With practice, critical thinking will become second nature. We should be careful how we apply it to everyday life as critical expressions of thoughts dur-ing casual conversations can be detrimental to personal relationships. As we become better critical thinkers, we become more discerning scientists, scientists who better understand the scientific literature and can find an appropriate niche to do research.

Food scientists also need to see a broader picture. We need to know where we fit into the world and to know what other people think of us. Processed foods, artificial ingredients, chemicals in our foods, and unnatural foods have long been the target of books (Davis, 1954; Gibbons, 1962; Turner, 1970; Dufty, 1975; Worwood, 1991; Nestle, 2002, 2006), but there seemed to be a limited understanding of food science. Three more recent books (Pollan, 2008, 2009; Kessler, 2009) directly challenge food science and all for which it stands. Food scientists have responded to the criticism of industrialized food (Floros et al., 2010). Another book (Paarlberg, 2010) seems to independently evaluate critics and supporters of the agricultural system, including the world of food science. A critical reading of one or more of these sources or alternatives is recommended to see if there is what we can learn as we practice our profession and how to better respond to criticism that we may receive.

Limits of Critical Thinking

Critical thinking is a buzz phrase used widely in universities across the country. Many universities are requiring a common book either for incoming First-Year students or for the entire undergraduate student body to build critical thinking skills. Sometimes it addresses a topic that is familiar to most students, provides an alternative view to the norm, and provides excellent interaction among students and faculty members. Too often it presents a point of view or agenda that a certain group of faculty wishes to inculcate onto unsuspecting students. Unfortunately some of these books are critical of processed foods or the food industry without any perspective on a scientific background. Critical thinking is not blindly submitting to someone else's opinion.

Critical thinking requires a knowledge base. To be critical thinkers, we need a degree of familiarity with the topic which requires time to research the area and understand the generally accepted premises associated with it. Thinking without knowledge leads to blind advocacy. At best, it can stimulate one to seek more knowledge and study of the arguments on all sides of the issue. At worst, it can lead to advocating extreme positions and even violence. Knowledge without thinking leads to blind following. At best, it can serve a noble cause. At worst, it can lead to cults and mass suicide. Although the development of critical-thinking skills is an important component in any undergraduate education program, development of a knowledge base is more important. In graduate education, all admitted students should have that knowledge base, and the development of critical thinking skills is crucial to the development of a research scientist.

Critical thinking is an essential component of any successful research program, but it is not an end in itself. It should help identify researchable objectives and place the research in context of the specific area of investigation. It can lead to perfectionism stymieing any research direction resulting in stimulating discussions but no research publications. As mentioned in Chap. 4, the successful scientist develops a synergistic relationship between the scientific literature and experimentation.

References

Afamasaga-Fuata'i K (2008) Students' conceptual understanding and critical thinking: a case for concept maps and vee-diagrams in mathematics problem solving. Aust Math Teach 64:8–17

Browne MN, Keeley SM (2011) Asking the right questions: a guide to critical thinking, 10th edn. Longman Publishing Group, New York

Burton R (2008) On being certain: believing you are right even when you are not. St. Martin's Press, New York

Davis A (1954) Let's eat right to stay fit. Harcourt, Brace, New York

Dewey J (1909) How we think. D.C. Heath and Company, USA

Dufty W (1975) Sugar blues. Chilton Book Co., Radnor, PA

Gibbons E (1962) Stalking the wild Asparagus. D. MacKay Co., New York

Grinnell F (1992) The scientific attitude, 2nd edn. Guilford Press, New York

Fisher A (2001) Critical thinking: an introduction. Cambridge University Press, Cambridge, UK

Floros J, Newsome R, Fisher W, Barbosa-C'anovas GV, Chen H, Dunne CP, German JB, Hall RL, Heldman DR, Karwe MV, Knabel SJ, Labuza TP, Lund DB, Newell-McLaughlin M, Robinson JL, Sebranek JG, Shewfelt RL, Tracy WF, Weaver CM, Ziegler GR (2010) Feeding the world today and tomorrow: the importance of food science and technology. An IFT scientific review. Comp Rev Food Sci Technol 9:572–599

Kessler D (2009) The end of overeating: taking control of the insatiable American appetite. Rodale Press, Emmaus, PA

McInerny DQ (2005) Being logical: a guide to good thinking. Random House, New York

Nestle M (2002) Food politics: how the food industry influences nutrition and health. University of California Press, Berkeley, CA

Nestle M (2006) What to eat. North Point Press, New York

Paarlberg R (2010) Food politics: what everyone needs to know. Oxford Univ. Press, New York

Paul RW, Elder L (2002) Critical thinking: tools for taking charge of your professional and personal life. FT Press, Upper Saddle River, NJ

Pollan M (2008) In defense of food: an eater's manifesto. Penguin Press, New York

Pollan M (2009) Food rules. Penguin Press, New York

Taleb NN (2010) The black swan: the impact of the highly improbable. Random House, New York

Thubron C (1978) Istanbul. Time-Life Books, Amsterdam

Turner JS (1970) The Chemical Feast, the Ralph Nader Study Group Report on Food P Protection and the Food and Drug Administration. Grossman Publishers, New York

West DC, Pomeroy JR, Park JK, Gerstenberger EA, Sandoval J (2000) Critical thinking in graduate medical education: a role for concept mapping assessment? JAMA 284:1105–1110

Worwood VA (1991) The complete book of essential oils and aromatherapy. New World Library, San Rafael, CA

Chapter 12
Science and Philosophy

A philosopher is a person who knows less and less about more and more, until he knows nothing about everything.

A scientist is a person who knows more and more about less and less, until he knows everything about nothing.

John Ziman (2010)

The scientific method has evolved over the years. Sir Francis Bacon, the first experimentalist may also have been the first food scientist. He used snow to freeze a chicken (Bolles, 1997). Initial research was verification. Investigators developed a hypothesis and then conducted an experiment to "prove" their hypothesis. If the experiment didn't work out, they needed to change their hypothesis and retest it.

At the beginning of the twentieth century, Hans Vaihinger [see translation from original German by Ogden (Vaihinger, 1965)] introduced the concept of "as if." In this phase, scientists developed assumptions which they treated "as if" they were true. While it was difficult to determine the truth, they could develop hypotheses that followed from the assumptions. Vaihinger called these bodies of assumptions "fictions." As long as the hypotheses were proven to be true, the fictions held. When hypotheses failed, new fictions needed to be developed. Primary examples of this type of investigation are taxonomy and atomic theory. As more information came in, organisms or atomic particles were reclassified or renamed. Microbial taxonomy still follows these principles.

The Uncertainty Principle was published in the 1927 by Werner Heisenberg (as described in Heisenberg, 2007). The principle comes from quantum mechanics and states that "The more precisely the position is determined the less precisely the momentum is known." A closely related principle that is frequently confused with the Uncertainty Principle is the Observer Effect which states that observation of an electron requires interaction, but interaction changes the path of the electron. If we expand these principles beyond the narrow interpretation in physics, it points out the uncertainty in any experimental protocol, the inability to truly observe something without altering the process and the subsequent interpretation of results.

R.L. Shewfelt, *Becoming a Food Scientist: To Graduate School and Beyond,*
DOI 10.1007/978-1-4614-3299-9_12, © Springer Science+Business Media New York 2012

Falsification was introduced in the 1930s by Karl Popper (as described in Popper, 1968). In falsification, multiple explanations/hypotheses are developed. Available information is then used to eliminate the least likely hypotheses and narrow the choices down to two—the null hypothesis and the alternate hypothesis. Experiments are then designed to reject the null hypothesis and accept the alternate hypothesis. Falsification applies when the results do not provide a clear "Yes/No" answer (Beveridge, 2004) such as a reaction either leads to an explosion or it does not lead to one. Falsification analyses are generally tested by statistical analysis. We do not really prove something in falsification. Rather we find evidence that corroborates our alternate hypothesis. Falsification provides "provisional conjectures" not "true statements."

Replacement was described in 1962 by Thomas Kuhn (and updated in Kuhn, 1996). Replacement involves "paradigms." A paradigm is a set of practices that help define a scientific discipline or area of research. It rests on a series of assumptions that are accepted by scientists in a discipline or research area. According to Kuhn, normal science involves "puzzle solving" which can also be described as hypothesis testing. During the course of experimentation, "anomalies" develop that don't make sense in the context of the paradigm. As they become apparent, anomalies are fitted into the existing paradigm. As it becomes more difficult to fit these anomalies into the paradigm, the paradigm needs to be revised. Eventually a new paradigm will be proposed leading to a scientific revolution. Rivalry will develop between new and old schools of thought. A true revolution occurs when the new paradigm replaces the old paradigm. On other occasions the old paradigm undergoes extensive revision, but there is no replacement. Rarely is there a compromise between the old and new schools, because the new paradigm is usually a radical departure from the old one. Kuhn describes the process in general and the pitfalls of both new- and old-school advocates.

An easy-to-understand example of a scientific revolution includes the center of the universe. Claudius Ptolomaeus (Ptolmey) claimed that the earth was the center of the universe. Nicolaus Copernicus initiated a scientific revolution changing the paradigm to the sun as the center of the universe. Albert Einstein started a new scientific revolution rejecting the idea that there is any center of the universe. Despite scientific evidence to the contrary, citizens of Massachusetts still believe that Boston is the center of the universe.

When I was in graduate school, I was investigating lipid oxidation in flounder muscle. During that time, there was a paradigm battle among fish muscle physiologists. Olaf Braekken (1956) had published a theory in *Nature* that red muscle in fish acted like a liver. Numerous articles were published on fish muscle physiology over the next two decades, and discussion became heated. In the end, the Braekken paradigm was rejected, but the intense scrutiny given to fish muscle and the attempts to disprove the liver theory greatly improved our understanding of muscle physiology in fish.

Another successful scientific revolution was in the area of philosophy of science as outlined above. Verification gave way to falsification which is giving way to

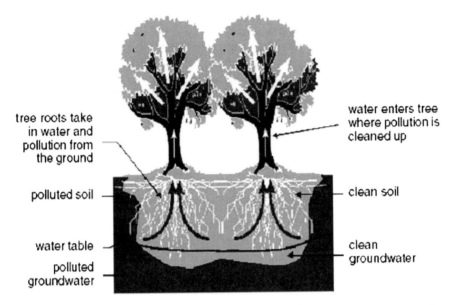

tree roots take
in water and
pollution from
the ground

water enters tree
where pollution is
cleaned up

polluted soil

clean soil

water table

polluted
groundwater

clean
groundwater

Fig. 12.1 An example of a scientific revolution in using plants to remediation of polluted soils. From *A Citizen's Guide to Phytoremediation* published by the Environmental Protection Agency

replacement, although many scientists today still see science as falsification. The most important perspective to us is the view of science by our major professor and the other members of our graduate committee. It is interesting to note that Kuhn, in his book, falls into all the pitfalls he warns new-school advocates to avoid.

Recent revolutions in science include the use of plants to clean up soil pollution (see Fig. 12.1) and the incorporation of pro-vitamin A into rice to improve nutritional quality of the most widely consumed staple in the world to form golden rice (see Fig. 12.2). Commercial success of this triumph in molecular biology to improve the nutritional quality of food will depend on consumer acceptability of the difference in color and subtle differences in flavor.

Most scientists don't worry much about philosophy. Many very effective scientists don't have an appreciation of the philosophy of science, they just "do" science. The implications from a philosophical perspective are that:

- "f**t" is a four-letter word
- Information does not need to be true to be useful
- There is no single scientific method

The practical implications of this chapter to beginning scientists are to:

- Adopt the prevailing paradigm
- Develop a personal thought style
- Conduct research within the paradigm
- Challenge the system thoughtfully and carefully

Fig. 12.2 Picture of golden rice which is grown to improve the nutritional quality of the grain. Courtesy Golden Rice Humanitarian Board. www.goldenrice.org

Forms of Reasoning

In general we divide reasoning into deductive and inductive reasoning. Deductive reasoning uses general principles to relate causes to effect. Inductive reasoning develops general principles based on detailed results. Deductive reasoning is important in developing hypotheses or research objectives. Inductive reasoning is important in relating our results to the remainder of the relevant literature. When writing a manuscript or reading an article, we understand that the results are only directly applicable to the conditions of the study conducted. The value of the research, however, is how well it can be generalized. Authors may be tempted to generalize their findings beyond credibility. We must be careful to view the general conclusions in our manuscripts and the articles we read critically.

Martin (2009) describes a third type of reasoning—abductive reasoning. Abductive reasoning is developing explanations for two or more possible contradictory sets of results or concepts. In science as in other areas of life, we are often posed with false choices—it must be either this way OR that way. Most of the time, science is more complex than an either/or decision. Abductive reasoning is particularly useful in developing deeper hypotheses or research objectives. It is also very useful in resolving anomalies that develop in a field of research. Abductive reasoners tend to be innovators in the field leading to breakthroughs. They also may be on the outside looking in if they fail to mold their thought styles to the thought collective as described below.

Thought Styles

A thought style is defined as a "set of assumptions enabling a scientist to observe and to act" (Grinnell, 1992). Most scientists cannot articulate their particular thought style. Many of them don't even realize they have a thought style, but all scientists have one. Thought styles assume that there is order in the universe and that we can learn that order through experimentation. Professors seek out students who can advance their thought styles. A thought style governs how a professor operates, thinks, and goes about daily business. It is how each scientist finds meaning in a "thought collective" which is described below. Scientists' legacies are established through their thought styles and their influence on their students. When a scientist makes an observation, forms hypotheses, or evaluates the prevailing paradigm it is in the context of the thought style. Science is not constant. What a scientist believes as scientific truth is part of the thought style, but beliefs can evolve with personal observations and published studies.

Thought styles determine what scientists look for and what they see. Thought styles are the prisms we use to observe and interpret data. Certain aspects of the data not obvious to the normal person may be noted and other aspects may be ignored. When Max Planck said that "Scientists never change their minds, but eventually they die" he was talking about the hold thought styles can have on a scientist. A thought style will affect how the laboratory is organized, graduate students and postdocs are recruited, technicians are hired, and what opportunities develop. Individual thought styles begin to form during graduate education. They are passed from one generation to another (a major professor to students, postdocs etc.). They determine the progress of everyone in the lab as they are evaluated in terms of the professor's thought style. Scientists also attempt to sell their personal thought styles to the thought collective.

A thought collective is defined as—"individual investigators interacting with each other share to some extent a collective thought style about how group activities should be carried out" (Grinnell, 1992). Thought collectives are important because it is the thought collective that decides who gets funded and what gets published. It determines the prevailing thought style and acceptable alternatives. When enough anomalies develop, the thought collective may undergo division into an old school and new school. Minor modifications to the thought collective may bring the two schools back together. If not, there may be years of scientific warfare until one of the schools prevails. If the old school wins, members of the new school must either submit or move on. If the new school wins, we have a scientific revolution and old-school scientists must learn to reinvent themselves or shut down their research and look for employment in teaching or administration.

Examples of Paradigm Shifts in Food Science

Several examples of paradigm shifts are available in many fields as described by Cohen (1985), Preston (2008) and Dear (2009). There are few such descriptions in food science. In the 19th century, American entrepreneurs Sylvester Graham, the

Table 12.1 Paradigms in food science education

	Instruction paradigm	Learning paradigm
Knowledge delivery	Lectures	Problem-solving and discussions
Learning	Accumulation of facts	Discovery of knowledge
Learning controlled by	Teachers	Students
Study	Individual	Teams
Success	Competitive	Cooperative
Instructors	Subject experts who primarily lecture	Course designers who empower student learning

Adapted from Iwaoka et al., 1996

Kellogg brothers and C.W. Post popularized the use of foods to promote health, wellness, and moral sufficiency, developing healthy flours and cereals that revolutionized food processing and manufacturing (Worthen, 2006). America has had a long history of health-food proponents and stores such as GNC (founded 1935; http://gnc.mediaroom.com/index.php?s=40), Earth Fare (started as "Dinner for Earth" in 1975 with a name change in 1993 http://www.earthfare.com/sitecore/content/EarthFare/OurCompany/History.aspx) and Whole Foods (founded 1980, http://www.wholefoodsmarket.com/company/history.php). Food scientists tended to look at these efforts skeptically until the 1989 when the term *nutraceuticals* was coined by the Foundation for Innovation in Medicine (Andlauer and Fürst, 2002). Food companies observed that there was money to be made in nutraceuticals. Food scientists followed by renaming the term *functional foods* (Schmidl and Labuza, 2000) with numerous books describing nutritional benefits (Arnoldi, 2004; Shibamoto et al., 2008), processing considerations (Mazza, 1998), and product development (Gibson and Williams, 2000).

Food science education is undergoing a paradigm shift from a delivery of knowledge in a traditional lecture and laboratory system to a more inquiry-based and discovery process (Iwaoka et al., 1996). Such innovations include use of journals, team-based learning, simulations, problem-based studies, and other techniques that more actively engage students in the learning process (see Table 12.1).

References

Andlauer W, Fürst P (2002) Nutraceuticals: a piece of history, present status and outlook. Food Res Int 35:171–176

Arnoldi A (2004) Functional foods, cardiovascular disease and diabetes. CRC Press, Boca Raton, FL

Beveridge WIB (2004) The art of scientific investigation. Blackburn Press, Caldwell, NJ

Bolles EB (1997) Galileo's commandment: an anthology of great science writing. W.H Freeman, New York

Braekken OR (1956) Function of the red muscle in fish. Nature 178:748–749

Cohen B (1985) Revolution in science. Harvard University Press, Cambridge, MA

Dear P (2009) Revolutionizing the sciences: European knowledge and its ambitions, 1500–1700, 2nd edn. Princeton University Press, Princeton, NJ

Gibson GR, Williams CM (eds) (2000) Functional foods: concept to product. CRC Press, Boca Raton, FL

Grinnell F (1992) The scientific attitude, 2nd edn. Guilford Press, New York

Heisenberg W (2007) Physics and philosophy: the revolution in modern science. Harper and Row, New York

Iwaoka W, Britten P, Dong FM (1996) The changing face of food science education. Trends Food Sci Technol 7:105–112

Kuhn TS (1996) The structure of scientific revolutions, 3rd edn. University Chicago Press, Chicago

Martin R (2009) The opposable mind: winning through integrative thinking. Harvard Business School Publishing, Boston, MA

Mazza G (1998) Functional foods: biochemical and processing aspects, volume 1 functional foods and nutraceuticals. CRC Press, Boca Raton, FL

Popper KR (1968) The logic of scientific discovery, 3rd edn. Hutchinson, London

Preston J (2008) The structure of scientific revolutions: a reader's guide. Continuum, Santa Barbara, CA

Schmidl MK, Labuza TP (2000) Essentials of functional foods. Aspen, Gaithersburg, MD

Shibamoto T, Kanzava K, Shahidi F, Ho C-T (eds) (2008) Functional food and health. American Chemical Soc, Washington, DC

Vaihinger H (1965) The philosophy of "as if": a system of theoretical, practical and religious fictions of mankind (translated from the 6th German edition by C.K. Ogden), 2nd edn. Routlege and K.Paul, London

Worthen DB (2006) The road to Wellville – the Kellogg story. Pharm Hist Australia 3(30):8–10

Ziman J (2010) Knowing everything about nothing: specialization and change in research careers. Cambridge University Press, New York

Chapter 13
Ethics in Science

People have to know whether or not their President is a crook.
Well I'm not a crook. I earned everything I've got.

Richard Nixon

It depends on what the meaning of the word 'is' is.

Bill Clinton

Most of us know right from wrong. Ethical practice is basically doing what is right and avoiding the temptation to resort to shortcuts, lies, cheating, and fraud. I write this chapter in trepidation fearing that rather than providing guidelines for appropriate behavior I may be providing temptation for bad behavior. As scientists we like to believe that we have higher standards than politicians. To be able to claim a higher calling we must not only practice ethics, but we must also be vigilant in holding colleagues and organizations to the same standard.

Data

As scientists, we are dedicated to the idea that data drive our conclusions. If it can't be demonstrated through a data-collection process it doesn't count. In science, nothing is sacred. All propositions must be testable. Science attempts to understand our world. If our hypotheses fail, we attempt to understand why, reformulate the hypothesis, and conduct more experiments. To be valid, the data must be true. Any manipulation of data defeats the purpose of science. Unfortunately, there are many ways to manipulate data that steer us away from really understanding what is happening.

Sindermann (2001) describes six ways that data can be manipulated:

- *Massaging*—transforming inconclusive data into apparently conclusive data
- *Extrapolating*—stretching a small amount of data beyond the limits of their applicability
- *Smoothing*—discarding data that does not appear to fit into preconceptions

R.L. Shewfelt, *Becoming a Food Scientist: To Graduate School and Beyond*,
DOI 10.1007/978-1-4614-3299-9_13, © Springer Science+Business Media New York 2012

- *Slanting*—emphasizing aspects of the data that support our conclusions while ignoring those that do not support them
- *Fudging*—creating data points that help us make our point(s)
- *Manufacturing*—creating mythical data sets that clearly prove our hypotheses

The degree of culpability increases from *massaging* to *manufacturing*, but all violations prevent us from adequately learning what is really happening. There are times where we can discard data. For example, if a power failure occurs in the middle of data collection, and the later samples are compromised, it may be necessary to discard all the data collected in the whole session. Use of an incorrectly prepared reagent or batch of media would invalidate all data from tests with that reagent or batch. Any clearly identified mistake constitutes a valid reason for discarding data. The point is to determine what the data are telling us NOT how to manipulate the data to tell us what we want to see. Note also that there are statistical techniques to determine outliers (Mason et al., 2003), but it is important to follow all of the rules and to state what we did in any reporting of the data.

Ideas

In science, we deal with ideas. Giving proper credit for an idea is part of the morality of a scientist. We shouldn't steal ideas from others. The scientific process is not always crisp and clear. It may be difficult to trace the evolution of every idea that we have had and to provide proper credit. It is even harder for those of us who are not obsessive about keeping accurate records of every presentation at a meeting, article we have read, or conversation we have had. Still, it is important that we find ways to give proper credit for our ideas.

Sindermann (2001) also describes four ways that we can treat ideas unethically:

- *Premature disclosure*—publicizing someone else's ideas before they are properly credited
- *Scientific ectoparasitism*—collecting other peoples ideas and developing then as our own (such as stealing an idea given in confidence from someone else and rushing out to write a grant based on that idea)
- *Mirror writing*—publishing another scientist's published ideas without proper citation
- *Plagiarism*—direct copying of words or data from other publications without credit

Once again the violations become more egregious as they proceed from *premature disclosure* to *plagiarism*, but the same warning holds. Any violation is a serous breach of ethics. We have an obligation to be fair in the exchange of ideas.

Ethical concerns have resulted in several scientific disciplines including physical chemistry. Rivalries between Arrhenius and Nernst, Nernst and Haber, as well as Langmuir and Lewis all went back to real or perceived stealing of ideas (Coffey, 2008).

Presentations and Manuscripts

There are some other things we must consider when making presentations at scientific meetings or writing manuscripts for scientific journals. Original data are considered unique and should only be presented officially once. When becoming part of the record, we are duplicating the credit we receive by presenting it again. For example,

- *Duplicate presentation of data*—presenting original data at one meeting and then turning around and presenting it at another meeting
- *Duplication of data in written form*—presenting original data in two separate publications
- *Primary data in review formats*—presenting primary data in a review article

There are some exceptions to these rules. If we present our data at a departmental seminar, that doesn't preclude us from presenting it at a regional or national meeting. A guideline we can use is whether there is a published abstract. Once the data has been featured in a published abstract, it should not be presented again as original. We want to be careful with the Internet. If our PowerPoint presentation at a departmental seminar ends up on the web, it then becomes official and is also fair game for other scientists to use those ideas (hopefully giving us proper credit!). Data are not to be duplicated in other articles. We present it once, and that is it. When writing a review article, we can present our data or that of others to make a point, but the original article must also be cited and we must obtain permission from the publisher. The publisher holds the copyright, not the author(s). A courtesy request from the corresponding author is not required but is recommended.

On the Job

Food scientists face ethical challenges on the job. Clark (2009) defines an ethical challenge as one that "requires judgment." He classifies these challenges into economic, interpersonal, or regulatory challenges. How we respond is the measure of our integrity.

Economic challenges can range from padding an expense account to taking bribes to falsifying records that benefit ourselves or the organization that employs us. Again, most of us know right from wrong, but is cheating either for ourselves or for the company really worth the chances of getting caught? Interpersonal challenges include gossiping, harassment, padding our résumé, and looking the other way when someone is obviously cheating. We have an obligation to treat others as we wish them to treat us. Regulatory challenges include falsifying records, misleading governmental agencies, selling dangerous products, and cover ups. How would we have handled the situation at the Peanut Corporation of America (see Table 13.1)?

Table 13.1 Timeline of events at Blakely plant of the Peanut Corporation of America (PCA) regarding peanut paste scandal adapted from American Institute of Baking (AIB) website (https://www.aibonline.org/press/AIBStatement04022009/Chronology.html) See also CDC (2009) and Jargon and Zhang (2009)

2005	PCA sales $15 million. AIB explains audit procedure to PCA Plant Manager.
2006	Georgia Department of Agriculture conducts four inspections and cites minor violations. AIB audits plant once and rates plant Excellent (score 875).
2007	Georgia Department of Agriculture conducts four inspections and cites minor violations. Three samples taken in August and test negative for *Salmonella* and pesticides. AIB audits plant once and rates plant Excellent (score 900). Technical Manager at the plant leaves late in the year.
2008	PCA sales $25 million. Seven tests show positive response for *Salmonella*. Tests of retested product are negative so suspect product is shipped. AIB audit in March rates score of 910 (Superior) but indicates some problems with maintenance and sanitation. Georgia Department of Agriculture inspects facility in June indicating that all violations have been corrected. Blakely Plant Manager leaves company in June. Centers for Disease Control and Prevention (CDC) reports illnesses due to *Salmonella* of unknown origin, September 8. Food and Drug Administration (FDA) joins CDC in investigating peanut products as a source of *Salmonella typhimurium* in December. The first death due to the outbreak occurs in Brainerd MN.
2009	CDC reports 388 consumers infected as of January 7. Later in the month, the source is identified as King Nut peanut butter produced by PCA in Blakely GA. PCA recalls 21 specific lots of peanut butter/paste. United States Department of Agriculture (USDA) bans PCA from all government contracts in February. PCA files for liquidation of assets. PCA employee is quoted in an *Atlanta Journal Constitution* article that the only time sanitation was conducted was in anticipation of a government inspection or AIB audit.

Clark (2009) provides some practical tests to determine if we are not sure whether we are making an ethical choice:

- Lying, stealing, and doing harm are almost always wrong!
- What would your mother think?
- Can you look yourself in the mirror?

I would add, "Can you sleep at nights?" One of the problems with making ethical choices is that the more we compromise our values, the easier it is to compromise ourselves the next time we face a challenge. See the examples below and factors in decision making in Fig. 13.1.

Some Hypothetical Ethical Challenges

You work for a Fortune 500 food company and develop a functional food that does well in market testing and have been invited to one of the marketing team's sessions to design the label and advertising campaign. The claims the team plans to make, while legal, are highly misleading. You express your reservations. The leader of the marketing team indicates that these claims are much more conservative than those

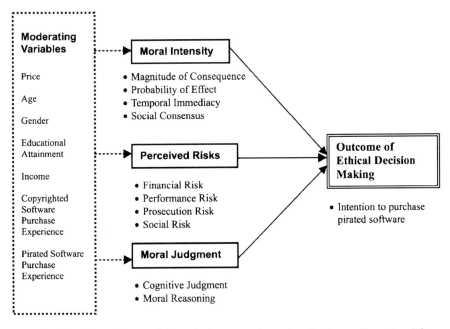

Fig. 13.1 Decision-making model on whether to purchase pirated software. Reproduced from Tan (2002) with permission from Emerald Insight

made by smaller companies with similar products that are distributed in upscale organic food markets. What is your responsibility as the only technical person in the room? Do you have any obligations to inform anyone else after you leave the meeting?

There is one special person in the lab who befriended you when you first arrived and helped you adapt to strange surroundings. He has become a valued mentor and a good friend. Lately you have noticed that he has been spending much less time in the lab due to some personal problems. Despite his absence, he still appears to be generating large datasets as evidenced by his presentations at group meetings. Knowing him and knowing the time required to conduct these experiments, you have a feeling that he might be manufacturing data. What are your options in this case? Who can you talk to about it?

You are conducting a sensory descriptive panel as a part of your dissertation research. You started out with twelve panelists, and most of them have been good panelists. You have three panelists who present problems. One, a professor on your committee, misses about half of the testing sessions due to schedule conflicts, but she is an acute taster and picks up on small subtleties that other panelists miss. Another, the technician in your lab, is always present, but his results are not consistent with those of the other panelists. Although your major professor believes that she is a keen sensory person, she can't detect bitterness in any of your samples, always overestimates sweetness and doesn't seem to understand the meaning of some key descriptors. Should you boot one or all of these individuals off your panel? Are there other options for you to consider? Should their data be discarded?

A student in another lab is a brilliant experimentalist. The two of you started in grad school the same semester. He typically spends 10–12 h in the lab a day generating data. His experiments are well-planned, and he is brilliant at performing statistical analyses. His problem is that his English is not good, and he has difficulty writing. He has heard you give presentations in class and read some of the team reports you have written for some of your classes together. He tells you that you are such a good writer and he needs your help. He offers a deal. He will do all of your statistical analyses if you will write his literature review. He has narrowed the articles he has read to 68 that should be cited in his literature review and has highlighted the key points in each article. All he needs from you is to string these thoughts into a credible literature review. You have tried to get help from your statistics professor on your data analysis, but she is not interested. How much writing can you do for him, and how much statistical assistance can you accept from him without crossing the ethical line?

One of the professors in your department appears to be stealing ideas from his students. He gives complex literature assignments in his graduate classes, pushing them to deeply analyze research articles. You found his class stimulating, and part of your growth as a graduate student is directly attributable to what you learned in the class. Last week, you learned that this professor just received a $500,000 grant on a topic that sounds remarkably similar to what you and your classmates were researching in his class. You wonder if the class was used by the professor to do his work for him. Is it legitimate for professors to use classroom assignments to provide material for grant proposals, journal manuscripts, or books they are writing? What can you or should you do when you see such cases?

Unfortunately, there are no easy answers to the questions posed above. Some of the scenarios posed above are more straightforward to you than the others. You might be surprised, however, when comparing your responses to those of your classmates. They might not have the same concerns about what their mother might think than what your mother might think. There are cultural differences and other reasons for different responses. All of these scenarios and situations in everyday life call for judgments. Most ethical challenges are not clear-cut. In most of them, there is a line that we must draw between right and wrong. It is where we draw that line that determines our character.

The most complete discussion on ethical behavior as a scientist has been written by Macarina (2005). For more ethical situations, see Seebauer and Barry (2001). For a broader approach to ethical considerations facing scientists today, I recommend Kurtz (2007) and Rollin (2006), which has a special emphasis on genetic modification.

References

CDC (2009) Multistate outbreak of *Salmonella* infections associated with peanut butter and peanut butter-containing products. MMWR 58 (Early release): 1–6. http://www.cdc.gov/mmwr/preview/mmwrhtml/mm58e0129a1.htm

Clark JP (2009) Ethical practices in the workplace. Annu. Mtg. Anaheim CA, Abst. 036–3. p 59. http://www.abstractsonline.com/Plan/ViewAbstract.aspx?sKey=5af4d21f-7320-47ac-abbf-b3c5a6ac86d6&cKey=fa3c4b73-78df-4f2a-8b88-2a65c8c187d9

Coffey P (2008) Cathedrals of science: the personalities and rivalries that made modern chemistry. Oxford University Press, New York

Jargon J, Zhang J (2009) Peanut-butter probe focuses on Georgia plant. Wall Street J, online. http://online.wsj.com/article/SB123194586477481479.html

Kurtz P (2007) Science and ethics: can science help us make wise moral judgments? Prometheus Books, Amherst, NY

Macarina FL (2005) Scientific integrity:Text and cases in responsible conduct of reserach, 3rd edn. ASM Press, Washington, DC

Mason RL, Gunst RF, Hess JL (2003) Statistical design and analysis of experiments, with applications to engineering and science. Wiley Interscience, Hoboken, NJ

Rollin BE (2006) Science and ethics. Cambridge University Press, New York

Seebauer EG, Barry RL (2001) Fundamentals of ethics for scientists and engineers. Oxford University Press, New York

Sindermann CJ (2001) Winning the games scientists play. Basic Books, Jackson, TN

Tan B (2002) Understanding consumer ethical decision making with respect to purchase of pirated software. J Cons Market 19:96–111

Chapter 14
Finding and Managing the Literature

Monica Pereira

> ... *[I]f we don't start seriously teaching our future users of information how to go beyond Google and benefit from more-sophisticated tools, we are helping to create a future in which if they can't find it with a search engine, they won't find it at all.*

> William Badke (2009)

In Chap. 4, we looked at evaluating the literature sources that we read. To make sure that we are finding all the relevant information, we need to develop our skills to search for those materials and organize them into a form that can be readily retrieved when needed. This chapter provides insight into how to develop those skills.

Organizing the Literature

Research projects rely on the appropriate use of relevant citations. Citations are references we cite in our papers and other projects because we have used ideas or quotations from them. Citations are culled from databases, catalogs, and other finding aids, called resources. Generally we find more references than we will actually use, but it is a good idea to collect comprehensively.

Complete citations contain all the elements that make a reference findable. For example, the following article citation:

Article Citation:
Klaauw NJ, van der Smith DV (1995). Taste quality profiles for fifteen organic and inorganic salts. Physiol Behav 58 (2):295–306

M. Pereira (✉)
Research & Instruction Librarian, Science Library,
The University of Georgia, Athens, GA, USA
e-mail: mpereira.lib@gmail.com

R.L. Shewfelt, *Becoming a Food Scientist: To Graduate School and Beyond*,
DOI 10.1007/978-1-4614-3299-9_14, © Springer Science+Business Media New York 2012

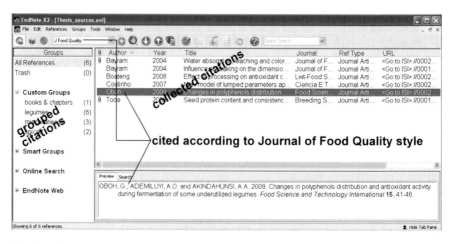

Fig. 14.1 Sample display from an EndNote library

contains *authors, year, article title, journal title, volume, issue,* and *page numbers.*
The citation was found by doing a topic search in the database, CAB Abstracts &
CAB Archives. While the citation might have been retrieved via some other resource,
in this instance a database was searched. Naturally other types of citations (e.g., book
chapters, patents, etc.) will require different elements when they are cited.

Read or scan the literature, keeping careful notes of where the various ideas or
quotations are located. It is crucial to keep a cache of pertinent citations in one place
rather than have them scattered among papers we have written, or among digital or
print scraps. Nothing is more distracting than to hunt for a misplaced citation or
article in the middle of writing a paper. For this reason, we should invest time in
learning to use a citation management system (CMS) like EndNote (see Fig. 14.1)
or RefWorks. While there are others, these two are reliable, and many academic
institutions make one or both of these available. Ask classmates or librarians to help
learn citation management efficiently and effectively. Using a citation manager to
create and maintain a database of citations is a process that will reap rewards over
the remainder of our career.

A CMS can store an unlimited number of citations, allow us to link to the full-
text, and add notes and commentary. It should be able to collect these citations from
various databases, and even the world wide web (Web). A CMS will also insert
selected citations within a paper and build a bibliography, according to a journal's
publishing style (e.g., APA). All the elements required for a citation will be assem-
bled according to the specified style. We may switch from one style to another on
the fly. If a style is not available, a CMS should allow us to edit an existing style to
suit, or create new ones.

If we are working with a group, having our citations in one place will speed up
the process of sharing. It may seem odd to begin a chapter on finding resources with
an overview of citation managers. However, if we are going to go berry picking we
need a bucket to hold our berries. The CMS is the bucket, and knowing how to use
it to store citations as we forage is simply good sense.

Understanding the Research Landscape

Pursuit of relevant literature must be done without preconceptions as to the out-comes, as mentioned in Chap. 4. Generic web searching can dazzle and mislead us into assuming that "everything", or everything worth reading, is on the Web because so much seems available when we run a search. Conversely, if we fail to find some-thing on the Web, we may overlook it because of its absence. Just because we think some information or an article *should* be online does not mean that it will be. Also, not everything on the Web is findable! The Web has many pockets that browsers cannot reach into. This deep (or invisible) web is not accessible to web crawlers. Furthermore, information may be protected by firewalls, require internal search engines, or be password-protected. Browsing the Web alone for research is a shal-low strategy because it cannot guarantee either consistency or comprehensiveness. Sacrificing these twin aspirations for the convenience of online access puts us in danger of building a superficial collection of sources.

Similarly, we cannot assume that everything can be found in any single database. Research exists in a wide array of formats, some of which can be ephemeral (e.g., face-to-face conversations, gray literature), and some that are highly processed (e.g., encyclopedias, handbooks). Make it a habit to search in more than one database or resource, and check the references of all the literature found. Information can take a number of formats. Figure 14.2 describes these formats in terms of the general ter-rain of information from the idea stage through tertiary resources.

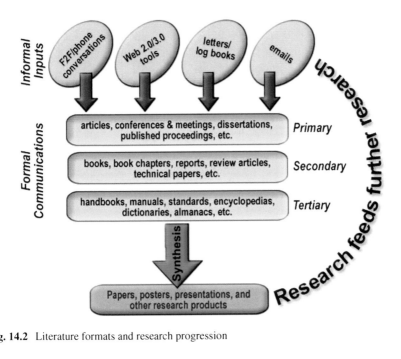

Fig. 14.2 Literature formats and research progression

Figure 14.2 also indicates that the process begins over again when new ideas are introduced. For example, conference paper presenters encounter further ideas when they share their paper, and these ideas are folded into an article. When other researchers read that article (among others) new or different ideas may be generated, and these in turn are disseminated, sometimes resulting in more publications, and sometimes, not. After all, not every paper we wrote as undergraduates grew up to be a published work! In fact, articles sent to a journal are invariability returned for revisions. Journal editors want to publish articles that promote the integrity of their journals. Still, the research process in every case is crucial to knowledge sharing; and many of these processes are iterative.

If we use Web resources (e.g., *FAOSTAT, List of Indirect Additives Used in Food Contact Substances*, etc.), we should capture their uniform resource locators (URLs) and verify them before citing them in case they have changed. Increasingly publishers use digital object identifiers (DOIs) to identify each electronic journal article. We should keep URLs and DOIs together with their references in our CMS to expedite access.

Reviewing the Literature

Research is inspired by ideas. Whether the ideas agree or argue against a position, or describe an experiment, they are not vacuum-sealed, but rather the result of thinking about existing knowledge, and building on it. We will find that literature reviews are a prominent part of published literature, especially in the sciences. The larger landscape needs to be defined, and our ideas situated within that landscape. This landscape is a strong motive to conduct comprehensive literature searches, in as many online and print resources as are available.

Researching the literature is a craft that requires knowledge and the patience to build skills that will serve us well. Use library catalogs and databases to find these items. If our library does not have titles, initiate a request from a cooperating library (if this is an option), or request them through interlibrary loan (ILL). Invariably getting theses and dissertations will mean using ILL. Finding out what library services are offered is a natural step on our research quest. There may be borrowing options, extended borrowing times, online loan renewal, and other services that are unique to our campus library. While it is not "wrong" to use generic Web browsers (e.g., GoogleScholar, Wikipedia, etc.), remember that whatever web-based material we find must be authoritative, and that convenience is never a deciding factor in establishing such authority. When using such sources, verify the information. When we find usable resources, we should study their bibliographies. We use bibliographies as signposts for furthering our research.

There are different classes of resources, as introduced in Chap. 4 and illustrated in Fig. 14.2. In the sciences, primary literature is considered original research, that is, designing, implementing, and reporting on experiments firsthand. These references are most often found in peer-reviewed journal articles, at conferences where researchers report on their findings in presentations and posters, and theses and dissertations. If the literature does not have to be peer-reviewed, food science and technology has a number of trade journals and other resources that provide useful information. Pictorial illustrations, anecdotal descriptions, and field reportage can fill in the gaps between the workplace and experimental functionalities of products and processes. They can also give working details of machinery and protocols, construction specifications, trade, and marketing tips that demonstrate how research is negotiated and applied in real-life situations.

Secondary literature tends to report on primary literature, or use the ideas of primary literature as if they were already known. This type of literature may be unpublished. For example, white papers, committee reports, and other documents created by local or regional committees may never be found except through serendipity. When published, secondary literature takes the form of the review article, which will contain a well-populated bibliography. Review articles are valuable because they collect and collate the relevant research for us.

Tertiary literature is far removed from primary research. This type of source will contain compilations of facts, syntheses, and overviews. The information will be presented as established, and we may not see sources for attribution of facts. The *CRC Handbook of Chemistry and Physics* is an example of a tertiary source in which the data is not sourced because it is assumed to be authoritative through countless iteration (e.g., melting/boiling points of substances). On the other hand, a tertiary source like an encyclopedia article will often provide a short bibliography. Also included in this level are handbooks and manuals. Now, on to finding literature for our CMS!

Finding the Literature: Where?

Now that we know what kinds of sources exist, the task will be to find the ones that are relevant to our research. Libraries generally have a range of databases in addition to their catalog. Browse the catalogs of other libraries. Be prepared to learn new ways to search. Not all catalogs and databases are equally easy to search. However, with a few tips, and the desire to explore their possibilities, we will be well-placed to conduct our research. The more comfortable we are with searching unfamiliar sources, the less spooked we will be when we encounter them.

Databases comprise rich wells from which to draw citations. Some of them are available through organizations like the Food and Agriculture Organization (UN), or the US Food and Drug Administration, so regardless of academic affiliation we can access them. Table 14.1 supplies a short list of such databases. Add to them as necessary. Depending on our research path, some of these organizations' publications may be influential in our careers.

Other databases may be restricted (password protected). We should make sure to know the password for our academic institution. Databases may be multidisciplinary (*Academic Search Complete*, *Web of Science*), or subject specific (*Food Science &*

Table 14.1 US federal and organizational websites

Current research information system (CRIS)	(http://cris.csrees.usda.gov/search.html)
The Federal Register	(http://www.gpo.gov/fdsys/browse/collection. action?collectionCode=FR)
Food & Agriculture Organizations of the United Nations Statistics Databases	(http://www.fao.org/corp/statistics/en/)
Food Safety & Inspection Service	(http://www.fsis.usda.gov/home/index.asp)
Food safety research information office (FSRIO)	(http://fsrio.nal.usda.gov/)
International bibliographic information on dietary supplements (IBIDS) database	(http://ods.od.nih.gov/health_information/ IBIDS.aspx)
Institute of Food Science & Technology	(http://www.ifst.org/)
United States Department of Agriculture	(http://www.usda.gov/)
United States Food & Drug Administration	(http://www.fda.gov/)

Caution: URLs may change

Table 14.2 List of potential databases and encyclopedias

Academic Search Complete
AGRICOLA
ASABE Technical Library
CAB Abstracts & CAB Archive
Compendex/Ei Engineering Village
Dissertation Abstracts
Food Science & Technology Abstracts
Food Science Source
INSPEC
Journal Citation Reports
Kirk-Othmer Encyclopedia of Chemical Technology
LexisNexis® Academic
LexisNexis® Congressional
MEDLINE®/PubMed®
Nutrition Abstracts
PolicyFile
Reaxys® (formerly Crossfire Commander/Beilstein–Gmelin)
ScienceDirect
SciFinder Scholar™
Scopus™
Specs & Standards (formerly HIS Standards; formerly TDX)
Ullmann's Encyclopedia of Industrial Chemistry
Web of Science/Web of Knowledge℠
WorldCat

Note: This list is not meant to be comprehensive. Talk to your librarian about relevant databases

Technology Abstracts).They may be designed to allow unique ways of searching. For example, *Reaxys®*, and *SciFinder Scholar™* will search for user-defined molecular structures, and *Scopus*, and *Web of Science* will perform cited reference searches. A useful list of databases is provided in Table 14.2. New databases may be introduced at any time. Keep current of what is available. We can use these databases to scan the tables of contents of specific journals, as suggested in Chap. 4.

Every article database will index the contents from a number of sources. Some sources may be dropped resulting in partial coverage of some titles, and new sources may be picked up. These decisions are made by the database vendors in consultation with the journal publishers. In time, there may be many more database options, so it is crucial to stay current with what our institution has, and what our colleagues at other institutions are using.

Since the number and type of resources can vary for each library, it is useful to know what resources our library has. Visit with a librarian to ensure we do not miss out on any resources! As mentioned earlier, our ILL department can usually provide for any inadequacies in a collection. At some institutions, this may come with a cost. Also, be aware of options that allow for user-direct access to resources. For example, state colleges and universities may belong to a consortium that allows us to initiate requests with a participating institution directly. Of course when the items arrive, we must visit our library to pick up the resources. As a rule, journal articles and databases may be off-limits to outside borrowers.

Finding the Literature: How?

Knowing how to search is a vital extension of knowing where to look. Electronic resources, like databases and catalogs, can make this process fast. With practice, we can make it efficient and effective.

Because journal articles are the mainstay of research at the university level, those are discussed first. When searching a database, we should turn the topic into a strategy by picking out the significant words, and combining them to achieve a result. Following is an example of a topic that has been operated on in this way:

TOPIC: Vegetable protetins are a viable substitute for meat proteins in the human
    diet.
STRATEGY: (vegetable* OR meat*) AND protein* AND (diet* OR food*)

A strategy is necessary because most databases cannot parse the meaning out of phrases and may use a wholesale algorithm which will include the words: *are*, *a*, *for*, *in*, and *the*. The resulting thousands of results will be frustrating to sift through. The challenge to picking out significant words rests on us. There will be obvious choices (vegetable, meat, proteins), and some less obvious ones, namely:

- Leaving out words (viable, substitute, human);
- Including alternate terms (food);
- Breaking up phrases like (vegetable proteins, meat proteins).

Note the use of parentheses to keep alternate terms together, like *vegetable* and *meat*. Alternate terms are separated by *OR* and kept within parentheses. Also note the truncation symbol, *, to ensure that all possible endings of a word are returned. *AND* combines a word or sets of alternate words such that the result will have at least one incidence of every word/word group combined using *AND*. The term for *AND* and *OR* used in this way is Boolean operator or logical operator. There is another Boolean operator that is used less frequently: *NOT*. An example of how to use it follows. Note that the entire search must be placed within parentheses when excluding a term or terms with *NOT*.

TOPIC: New processing methods for fruit and vegetable juices.
STRATEGY: ((fruit* OR vegetable*) AND juice*) NOT (thermal* OR heat*)

This example demonstrates that we can eliminate alternate terms as well. A less efficient search would have been:

STRATEGY: "new processing method*" AND (fruit* OR vegetable*)

Forcing the phrase *"new processing method*"* can eliminate some potentially useful results. Elimination of terms can be hazardous to our search. If one of those terms is used as contrast to another, those results will be eliminated. What is worse, we will not even know anything is missing. It is recommended that searches be conducted by matching *AND* and *OR* to what words are desired in the results. As a final tactic, *NOT* may be used to eliminate the obvious irrelevant terms.

Unique Database Features

Some databases offer unique features. *SciFinder Scholar* allows us to draw molecular structures (and save them), a useful way of identifying and searching for particular compounds. It also includes patents (US and worldwide) of relevance to food science. CAS Registry Numbers and references to spectral data are also available, as is the ability to limit searches to analytical aspects, substance preparation, and many other facets.

Specs & Standards is a database of official standards endorsed by official standard-making organizations worldwide. Since these documents are expensive, the chances are high that we will have to get them through ILL or purchase them because no library will be able to pay for all the standards to be available.

Scopus and *Web of Science* allow cited reference searching. This is useful for uncovering resources that have cited the ones we have already collected. Naturally, the more a citation has been cited by other researchers, the higher its usefulness quotient. Following this trail can often lead to some rare and useful discoveries. Also, the obvious seminal articles can easily be identified.

Journal Citation Reports, mentioned in Chap. 4, will provide journal rankings by their impact factors. Impact factors indicate which journals tend to have highly cited articles. The implication is that many researchers use those journals.

Catalogs

Catalogs are special databases that contain the holdings (monographs, journals, etc.) of individual library collections, or sets of many libraries' collections. They list what is available at college and university libraries, or in a special library collection.

The value of catalogs is often underrated. An online catalog will collate a library's collection by subject, assigning similar base call numbers (usually Library of Congress call numbers rather than Dewey Decimal ones) for a topic, and then accounting for the specific Library of Congress subject contents of a book by using more letters and numbers.

Think of this system as a code to a library's collection. Become familiar with the call numbers that pertain to our areas of interest to provide us with a shortcut to finding materials on the new bookshelf, resorting shelves, or in the stacks. If our primary institution's library does not carry an item, do not forget to request them through ILL. Most academic libraries, some special collections, and public libraries too, are partners in this cooperative lending venture.

Social Media: New Aids for Research

A 21st-century chapter on finding research literature would not be complete without mentioning the power of social media. Information used to flow along predictable print-based paths as if it were a river in a well-established riverbed. With the advent of the Web, that river of information became a flood. The Web offers evermore interactive ways to collaborate with colleagues across state and national boundaries. Research is driven by the need to share data and information in real time, and receive feedback as quickly as telecommunications will allow.

Social media are electronic outlets that include: blogs (weblogs), content-specific communities, forums, microblogs, news/RSS feeds, photo sharing, social networks, taxonomies, twitterfeeds, video sharing, virtual worlds, wikis, and other inventive avenues for sharing. Social media (sometimes called Web 2.0) are compelling because they are easy to use, centered on user interests, encourage input, and invite feedback and commentary. Their use, while not initially designed with academics in mind, has nevertheless been adopted by researchers as a mechanism for collecting and disseminating information.

RSS feeds are used for news alerts or microblogs (e.g., Twitter). Blogs are beneficial for discussing purposes, providing an arena for cultivating and testing ideas. Taxonomies (also called folksonomies) are used to arrange and relate concepts in ways that can highlight new conceptual relationships. Wikis (or other site-defined applications, like Google Docs) can be used to share time-sensitive information for a specified group of users. For example, a lab can set up a wiki for laboratory protocol, meeting minutes, and other shared documentation. Photo- and video sharing are used to illustrate, demonstrate, and even garner interest in, a variety of procedures. Audiovisual material can assist in demonstrating laboratory methods, experiment processes and results, like electron microscopy. Check relevant organizational websites to see which of these social media are available. For example, the FDA issues alerts using Facebook, blogs, and RSS feeds. In short, the flexibility of social media tools has allowed researchers to stay current and collaborate in ways that spark human creativity and afford new ways to encounter, capture, and use information. Remember that these media are just tools, and not the only tools we will use in our research adventures.

In conclusion, unless we commit to being comprehensive in our research, thinking about it does not amount to anything! This chapter provides an overview of the small processes that are intrinsic parts of the larger endeavor of producing research output. I recommend that we

- Have a place to put our citations so they will not be lost
- Know where to look for different subject information, and be open to new resources
- Know how to search for information, using logical operators, date each visit, and revisit them every few months for additions to the literature
- Be open to new ways of collecting and sharing information through social media

Remember to backup files regularly so we do not have to recreate previous steps in our research. Backup our CMS collection. Backup our search strategies and which resources we used. Backup any writing we do. Organize the files so we can tell which backups are the most recent. Resist the urge to tidy up by throwing away any digital files until projects are completed or published. Virtual copies take up so little physical space, and they are our insurance for peace of mind in our research process.

Reference

Badke W (2009) How we failed the net generation. Online 33(4):47–49

Chapter 15
Planning

Success is a lousy teacher.

Bill Gates

There are two kinds of failures: those who thought and never did, and those who did and never thought.

Lawrence J. Peter

Strategic planning differs from other types of planning in that it is broad in scope, starts with a mission or long-range goals, and develops strategies to achieve those goals. There are many definitions available, most of which deal with organizations, but I prefer the definition advanced at http://www.businessdictionary.com/definition/strategic-planning.html

> systematic process of envisioning a desired future, and translating this vision into broadly defined goals and objectives and a sequence of steps to achieve them.

Unlike typical long-range planning, strategic planning begins with the end in mind and is only useful if it is coupled with effective implementation. This definition permits strategic planning for a laboratory or even an individual. In a sense, all effective planning activities should be strategic in nature. Typically strategic planning involves a SWOT analysis which identifies the strengths, weaknesses, opportunities, and threats of the situation to maximize the chances of success. Strengths and weaknesses are generally considered to be internal to the organization and opportunities and threats are considered external. Effective planning is critical to success of any individual or organization.

Planning for the Oral Exam and Defense

One of the scariest and most important days in the life of a graduate student is the oral exam and defense of the thesis or dissertation. It marks a critical juncture in the student's program and career. Different schools handle written comprehensives

R.L. Shewfelt, *Becoming a Food Scientist: To Graduate School and Beyond*, DOI 10.1007/978-1-4614-3299-9_15, © Springer Science+Business Media New York 2012

differently, but at least once in every degree there is the dreaded oral exam of the student generally by the three to five professors who comprise the student's committee. Some careful planning might be useful. I have observed that there tends to be at least five purposes of oral exams to

- Determine if the students are qualified to get the degree
- Critique the appropriateness of the writing, illustrations, and references in the prospectus, thesis, or dissertation that has been presented
- Evaluate the soundness of the science that has been presented in the document being evaluated
- Teach students that they do not know as much as they think they do
- Teach students that they know more than they think they do

To prepare for this onslaught, we should consider doing the following:

- Review the key concepts of different areas of food science, particularly those taught in the classes taught by committee members
- Be prepared to defend any concept described in the text of the prospectus, thesis, or dissertation
- Know the basic principles of any instrument or method described and the reason why it was the most appropriate method for these experiments, even if it was chosen by the major professor
- Be able to present alternate conclusions to those drawn and to defend those that were drawn as the preferred ones
- Be able to illustrate specific concepts, draw structures of important molecules, or sketch pictures of key microbes
- Read about news events in the last month that have relevance to our research

RULE # 11

If you don't know the answer to a question, say you don't know. If you are speculating, say you are speculating. (told to me by my major professor Dr. Esam S. Ahmed the day before I defended my MS thesis).

It is important to listen carefully to the questions asked and answer each specific question. If unable to answer a specific question, it is better to plead ignorance than to wing it and be wrong. A student who can answer no questions will fail. A student able to answer all of the questions has a poor excuse for a committee. One trick a committee uses is to determine the level of understanding of concepts starting with rather elementary aspects and probing deeper until the student can no longer answer the question. Frequently the student will be led to three or four loose ends and then asked some questions that help tie these loose ends together. It is also important never to directly challenge a committee member. Remember they are the examiners and not the examined. We should not hesitate, though, to put forth our perspective based on our results or to cite certain labs that provide alternate perspectives. Finally, we should always be able to describe the most significant finding in the research; the relevance of the research, if any, to the food industry; how we could communicate the importance of our work to the public; and what we would do differently if we had to do it all over again.

Project Planning

In graduate school, we may or may not have the opportunity to intensively plan the overall project that encompasses our research. The way many students are funded is through a grant received by their major professor. The general objectives may have been established and the methodology outlined. Specific aspects of our graduate research must be developed. Some graduate students are given great leeway to plan specifics. Others are restricted to a plan developed by the major professor or a post-doc in the laboratory.

Upon graduation with a "real job," we will be expected to develop projects. In some cases, we may be given long-range goals and be expected to develop the action plan. In other cases, we will be expected to set long-term goals and develop the whole program. We may be expected to develop and submit our project plan, and there may be a specific form to complete. In other cases, there may be no requirements or forms to complete.

The first step in project planning is to develop goals, possibly in the context of the organization's mission statement. Some cautionary notes on mission statements:

- Some represent the guiding principle for everything that is done in that organization
- Some represent wishful thinking but are hopelessly out of date
- Some may be so broad that they are practically meaningless
- Some are in the process of being reevaluated to become relevant again
- Others may be buried away somewhere in the organization's documents and impossible to track down

Before extensive planning, find out the status of the organization's mission statement and its strategic plan if there is one. Our lab is our organization while in graduate school. Few labs have mission statements, but the department probably does. As graduate students we can comfortably ignore our department's mission statement, but our major professor's thought style will direct the mission of the lab. Beyond graduation, mission statements may become more relevant.

The definition of long range will vary greatly by organization. In graduate school, long range is between now and graduation. In some product-development situations six months or less is long range. For an assistant professor, long range is probably best defined as the time between hiring and tenure. Others may consider long range to be a career with a possible goal of a Nobel Prize. For each long-range goal, develop specific objectives. Remember the rule on objectives.

They should be clear, brief, achievable, and consistent with our goal. Clearly state the rationale for our overall project—what we hope to achieve and why is it important. Clearly state the problem and how to solve it. Next, outline the literature that provides a pathway to the objectives and the current understanding of the subject. Literature investigation will always be a work in progress that could result in modification of specific objectives and even long-term goals. Likewise, identify the procedures needed to conduct the individual studies which are also subject to modification as the project develops. In addition, identify the budget needed, how to obtain it, and a plan for achieving the objectives and goals. In an academic setting, grant writing will be part of the strategic planning process. Careful prioritization of the goals is critical for success. Generally speaking, the steps most likely to be successful and garner credit within our organization should be linked directly to our long-range plan.

A long-range plan should be dynamic, not static. It should be available for periodic review and perhaps modification. A plan that is never viewed has little value beyond the initial push that it gives. A plan that is never modified will soon become obsolete. A plan that is modified weekly or monthly is not a long-range plan. A good plan is one that is rigid enough to provide direction and prevent aimless wandering off track but flexible enough to remain relevant to changing circumstances.

Interdisciplinary Research Planning

A discipline was defined in Chap. 1 as a field of study. Informally we consider academic departments as disciplines, but formally we consider the more basic sciences as disciplines and the more applied sciences as applying the disciplines to specific topics. Thus, informally Food Science is considered as a discipline, but formally the disciplines we apply are primarily biochemistry, chemistry, microbiology, and engineering but also include physics, psychology, and biology.

I had the opportunity to communicate with Dr. Glenn W. Burton who was perhaps one of the two most famous research scientists our college ever produced. He was well-known for his cooperative research and sent me his model requirements shown in Fig. 15.1. For more information on the life and contributions of Dr. Glenn W. Burton, see his obituary at http://www.washingtonpost.com/wp-dyn/content/article/2005/11/24/AR2005112400936.html.

Many food scientists are content to work within their discipline or specialty area. Many research problems today are not clearly confined to a single disciplinary area. In response to these problems, scientists develop cooperative research teams to tackle these difficult problems. There are many forms of cooperative research. Sometimes a scientist is needed to fulfill a service function such as an expensive piece of equipment that is housed in the laboratory. This collaboration could include access to the equipment for us with grant funds and publications for the collaborator, but a scientist does not develop a noteworthy reputation based on service activities alone.

REQUIREMENTS OF EFFECTIVE COOPERATIVE RESEARCH

- **a leader**, someone who takes charge and is willing to make hard decisions,
- **mutual interest**, everyone on the team should have a passion for the subject and be highly motivated,
- **mutual planning**, including the ability to ask good questions and the willingness to learn about other disciplines/ specialties,
- **proper attitude** among all members who are willing to listen to different perspectives,
- **agreed responsibilities** that are written down and mutually agreed upon,
- **work**, everyone must pull their own weight,
- **publication that recognizes all cooperators**, or there will be dissension.

Fig. 15.1 Dr. Glenn W. Burton's recommendations for effective cooperative research

Another type of research cooperation is that of shared interests. For example, a food microbiologist and a food engineer might get together to improve a type of food process that requires less damage to quality while maintaining the safety of the product. Each investigator works at the interface of the two areas, but neither scientist directly crosses the line into the other specialty. Interdisciplinary teams involve three or more scientists focused on a specific problem. They can be working at the interfaces of a problem in which a chemist, nutritionist, and sensory scientist work to design a functional food that is nutritious with a highly desirable flavor. The three scientists do not necessarily need to know that much about the other areas as long as they can contribute their expertise when needed. Another approach is a division of responsibilities where each scientist is assigned a particular set of tasks to perform, some of which might go outside the individual's area of specialty. A third type of interdisciplinary research involves a more integrated approach where every team member must know at least a little about the disciplines/specialties of other members.

There are many benefits of interdisciplinary research. A truly integrated approach to interdisciplinary research expands our perspective as scientists. We begin to see that other disciplines/specialties have different ways of thinking that are not necessarily better or worse than our perspective, just different. Scientists tend to get so narrow in their perspective that they fail to see the broader picture (Ziman, 2010). In an integrated team, we are forced to have a clearer image of the broader picture. Interdisciplinary research provides broad-based solutions to problems. Narrowly focused research can lead to more problems than it solves because of unintended consequences or ignorance of key aspects of a problem at the broader level. Integrated-interdisciplinary research forces each investigator to address those problems that are usually ignored by more narrowly focused studies. More agencies are now providing funding support for interdisciplinary efforts.

Some grant programs actually require an interdisciplinary approach before considering a proposal. Interdisciplinary research generally provides a scientist's research more exposure to more scientists than a single subspecialty. It also provides an opportunity for situational leadership in which one scientist will assume

An example of an interdisciplinary team included an agricultural engineer specializing in handling systems, an extension food scientist specializing in fruit and vegetable processing and me, a food scientist specializing in fruit and vegetable quality. We were later joined by an agricultural economist interested in fresh fruits and vegetables. The economist urged them to abandon processed products and focus on fresh items resulting in the departure of the extension food scientist and the addition of a horticulturist specializing in soil science as it related to vegetables. Later an agricultural engineer specializing in mathematical modeling and a horticulturist specializing in preharvest cultural treatments of vegetables became members of the team. Many other scientists were cooperators on allied projects. One symbol of our work was a mobile laboratory shown in Fig. 15.3 which we could take to fields and packinghouses. The cooperation resulted in numerous publications including a book (Shewfelt and Prussia, 1993), a USDA National Superior Service Award (1988), and the start of an international conference on Postharvest Handling of Fresh Fruits and Vegetables. A combination of factors, not all of them pretty, led to the dissolution of the team, but the ideas lived on and subsequent books resulted from the earlier contributions (Shewfelt and Brückner, 2000, Florkowski et al., 2000, 2009).

Fig. 15.2 A true-life story of the postharvest systems team at the University of Georgia

leadership of one phase of the operation and others will serve as leader in other areas. Dr. Burton's model of a single leader was very effective for him and his collaborators on turfgrass research. Situational leadership worked reasonably well for the Postharvest Systems Team described in Fig. 15.2, but its team were not as successful and did not last as long as Dr. Burton's team.

There are many limitations to and frustrations associated with interdisciplinary research. First, teams need to deal with multiple personalities. If we have frustrations with members on team projects in our classes, multiply those frustrations by about ten. Different team members may not want to pull what we assume to be their weight, and they may have very different ideas of what needs to be done. That perspective is now coupled with scientists with big egos and little or no appreciation for or interest in food science! Communication is one of the biggest problems. Each department/discipline has its own jargon that is not always easily understood by others, but it is not the different terms that cause the most problems because we can always ask for a definition. The terms that are similar in both fields but mean different things are the ones that get in the way. The way *quality* is defined by a horticulturist differs from that of an engineer, an economist, and a food scientist. If we assume that everyone else has the same concept for the meaning of specific terms, we can bring much grief on ourselves. Beyond terms, different members have different philosophies (or thought styles) that are influenced by their specific disciplines. For example, food scientists think about preservatives, organic foods, and fortified foods differently than the general public. Many scientists we deal with are more likely to hold concepts similar to those of the general public. Imagine a typical food scientist working on a project with an ecologist! Likewise, food scientists may hold some strange ideas held by the general public that do not fit into another discipline's dogma.

A final problem associated with interdisciplinary research is bureaucracy. Even two labs right next to each other may have problems as to who pays for glassware, chemicals, and supplies needed for a joint project and who can operate specific instruments. Misunderstandings can lead to problems under the best of circumstances, but those problems are magnified when performing interdisciplinary research. Almost all Department Heads will encourage interdisciplinary projects, but they do not usually like to see departmental funds going to a program outside the department. For example, a food scientist was collaborating with a horticulturist who was looking for a Ph.D. student to work on a project that is primarily a food science project. The horticulturist had full funding for that student but no students in the department were interested. The food scientist could have provided a student, but for the student to get the money, the Ph.D. degree needed to be in Horticulture, not Food Science. Most food science students were reluctant to get their degree from another department, but finally one crossed the line.

When grant funds come into a university, they usually go to a specific department and it is sometimes difficult to work out the logistics of spending money across two or more departments. In addition, it is sometimes difficult to properly apportion credit. If an interdisciplinary team is small, all investigators tend to receive reasonably equal credit for the innovative nature of the team and numerous publications. As the team matures, however, and the membership changes, newer members may be regarded as providing service for the original members and not providing leadership, a key component of a scientist's reputation (Rosei and Johnston, 2006).

An interdisciplinary team can benefit from different preferred learning styles described in Chap. 1. Assimilators read broadly and can see linkages that others do not see, but they can be too theoretical. Convergers think outside the proverbial box bringing fresh ideas some of which can be brilliant, but many of these ideas can distract the team from its main goal. Divergers are good at formulating experiments, generating data, and analyzing it, but they may design experiments that are more interesting to them than relevant to the research goal. Accommodators are excellent publicizers, converting the knowledge to usable product, but they tend not to be detail oriented. Taking advantage of the strengths of each learning style while minimizing the weaknesses is a key to success. The Postharvest Systems Team had classic examples of an assimilator, a converger, and an accommodator. The Converger developed the overall concepts and sold them to the Assimilator and Accomodator. The Accomodator steered the project to more relevant projects. The Assimilator turned Diverger to develop the experimental treatments. There were many disagreements between the three until the data started rolling in and all investigators were so busy writing journal articles they did not have time to argue. The Accomodator suggested that a book be developed and negotiated the contract. All three collaborated on developing models to explain the data, and all three rode to tenure primarily on the basis of their membership in and contribution to the Postharvest Systems Team (Fig. 15.3). More details on interdisciplinary research as it pertains to research funding are provided in Chap. 16.

If you don't know where you are going, you might end up someplace else.

Yogi Berra

Fig. 15.3 Picture of the postharvest systems team beside the mobile lab which became its symbol

Career Planning

It is hard to think about career planning when our emphasis is on studying for the next test or completing the next experiment, but to move ahead, one must think ahead. Career planning is important because it

- Provides meaning to work
- Provides focus to enable progress
- Helps identify needs
- Gives us a measure of progress

RULE # 12
A scientist's career doesn't start with the first job. It starts with the first experiment that scientist plans.

In planning a career, some questions that should be answered include:

- What do we want to accomplish in a career?
- What are we willing to sacrifice to succeed?
- What are our specific goals?
- What do we need to do in the next year, 2 years, 5 years, and 10 years to reach our goals?

One way to proceed is to write down goals, sacrifices and steps to get there to be saved in a place for reevaluation at least once a year.

A CAREER FABLE

Once upon a time there was a visiting scientist who talked to a reasonably successful scientist one evening when both of them were taking a break. The visitor's goal was to become a world-renown scientist in her field. Unfortunately she wasstuck in a position under a Professor in a system that would not allow her to move ahead. The scientist suggested that she look for an appointment elsewhere. The problem with that suggestion was that she would have to relocate far away from her mother andbrother. She had some tough decisions to make. Time was running out for her professionally. If she did not make a move in the next two years, she was probably not going to achieve her goal, but that would probably mean leaving her home country and moving away from her family.

Fig. 15.4 Frequently, we need to make difficult decisions to advance our career

There are many factors that will affect a career plan. For example, identify:

- Personal strengths and weaknesses
- Ways to best exploit those strengths
- Ways to overcome or work around weaknesses
- Objectives needed to meet to achieve the career goal
- Restrictions on the job in meeting the career goal
- How much academic training is needed to achieve that goal

In today's climate, education is key. Will an MS degree be enough or will it take a Ph.D.? Maybe a postdoc would also be a good idea. Should these degrees all be at one school or spread over multiple schools? If looking at a MS degree, what are the advantages and disadvantages of skipping the MS degree and going straight to a Ph.D.? When sorting out the type of job upon graduation, some important questions include

- What type of facilities will be needed?
- Would the best location be at a university, in industry, in government, or self-employed? Or will it require a mix of locations?
- Would a big setting or a small one be preferable?
- Is independence more important than working in collaboration? How can the most appropriate collaborators be found?
- How will the projects be funded?

Two fables that are based on true stories are shown in Figs. 15.4 and 15.5. More details on career development are presented in Chap. 18.

Life Planning

When a food scientist was having a mid-life crisis, she heard a talk by Dr. Dave Lineback, a former President of IFT, who talked about developing a strategic plan for her life. She took the talk to heart and developed a personal strategic plan. She

A LIFE FABLE

Many years ago a little boy decided that he was going to get a Ph.D. in some scientific field. It was in his freshman year during his first food science course he made a commitment to becoming a food scientist. Pursuit of his goal was delayed by four years to serve his country in the military during the Vietnam era. When he became a graduate student he had two dreams – (1) teaching in a Food Science Department at a major university OR (2) managing a research institute where he would assemble a group of highly qualified scientists aimed at solving specific problems. It took him almost seven years to complete M.S. and Ph.D. degrees in food science. Many unproductive days of research led to the changing of his career objective to numerous occupations including a beach bum. He eventually ended up at a major university spending half of his career in a 100% research position and the other half in a predominantly teaching position.

Fig. 15.5 Some dreams come true while others fall by the wayside

identified seven critical objectives based on the IFT strategic planning process and developed an action plan for her life and career. Two specific objectives involved a career shift from research to teaching and a stronger financial basis for her retirement. Within three years, she made her desired career move and her financial situation is looking much better for a timely retirement.

Any life plan should place the career in terms of family obligations, financial considerations, outside hobbies, spiritual journeys, mentoring, and giving back to the community. Some people seem unable to make any contributions; others achieve great things. Organization and planning make a difference. Two books having helped me in my planning efforts. Covey (2004) introduces principle-centered leadership, writing about paradigms, and working within a paradigm. He emphasizes the difference between effectiveness and efficiency. He indicates that efficient is doing things right, but effectiveness is doing the right thing. Too often we become efficient without becoming effective. Cohen (2005) approaches the subject from the opposite direction describing how we can set ourselves up for failure by not thinking through our daily actions and activities. There are numerous other books (such as Kaufman, 1991), tapes, and other plans to help us organize our lives. Accessing a few of these in our spare time to glean ideas might help in developing a personal strategic plan.

References

Cohen A (2005) Why your life sucks and what you can do about it. Bantam Books, New York
Covey SR (2004) The seven habits of highly effective people, 2nd edn. Free Press, New York
Florkowski WJ, Prussia SE, Shewfelt RL (2000) An integrated view of fruit and vegetable quality. Technomic Press, Lancaster, PA
Florkowski WJ, Shewfelt RL, Brückner B, Prussia SE (2009) Postharvest handling: a systems approach, 2nd edn. Academic Press, San Diego, CA

Kaufman RA (1991) The strategic planning plus – An organizational guide. Scott Forseman Professional Books, Glenview, IL

Rosei F, Johnston T (2006) Survival skills for scientists. Imperial College Press, London

Shewfelt RL, Brückner B (2000) Fruit and vegetable quality: an integrated view. CRC Press, Boca Raton, FL

Shewfelt RL, Prussia SE (1993) Postharvest handling: a systems approach. Academic Press, San Diego, CA

Ziman JM (2010) Knowing everything about nothing: specialization and change in research careers. Cambridge University Press, Cambridge, UK

Chapter 16
Grantsmanship

There is no success without hardship.

Sophocles

It is not enough to succeed. Others must fail.

Lawrence J. Peter

Grant Writing

Grants are the lifeblood of any scientific investigator at a university. It is almost impossible for an Assistant Professor with a research appointment in Food Science to be promoted and tenured without obtaining grants. Grants are needed to purchase new equipment, fund graduate students, and grind out publications. The politics of obtaining funds from industry and government are somewhat different, but both require justification and written reports to obtain the necessary monetary support to run a lab. In industry, funding tends to be directed at reducing operating expenses or generating profit for the company although some companies still retain a basic research component that serves as a long-range incubator of product ideas and a status symbol. Government tends to operate more like universities but with a more rigid structure of project proposals, periodic reporting, and performance review. The rest of the chapter will be devoted to grant writing by university researchers, but both business entities and governmental researchers may be involved in pursuing federal grants. Industry scientists may never write a formal grant proposal, but they may be required to serve as liaison to a university project being funded by their company. Likewise, a governmental scientist may be asked to review grant proposals by various agencies.

The first step in any grant process is finding a funding source that matches our research interests and capabilities with their program objectives. There are numerous funding sources looking for scientists to meet their needs. An intensive search is usually necessary (Chapin, 2004). Most universities have searchable databases by keyword. They may also have an alert system for announcements of new projects

R.L. Shewfelt, *Becoming a Food Scientist: To Graduate School and Beyond,*
DOI 10.1007/978-1-4614-3299-9_16, © Springer Science+Business Media New York 2012

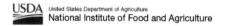

Agriculture and Food Research Initiative Competitive Grants Program

The purpose of AFRI is to support research, education, and extension work by awarding grants that address key problems of national, regional, and multi-state importance in sustaining all components of agriculture. AFRI supports work in six priority areas: plant health and production and plant products; animal health and production and animal products; food safety, nutrition, and health; renewable energy, natural resources, and environment; agriculture systems and technology; and agriculture economics and rural communities. In FY 2011, AFRI will support work in the six AFRI priority areas in a Foundational Program RFA to continue building a foundation of knowledge critical for solving current and future societal challenges. Additional RFAs further address AFRI priority areas in five societal challenge areas: Childhood Obesity Prevention; Climate Change; Food Safety; Global Food Security; and Sustainable Bioenergy. Funding opportunities for pre- and postdoctoral fellowship grants will be offered in a single, separate RFA.

Fig. 16.1 Overview of the AFRI program extracted from the NIFA website http://www.csrees. usda.gov/fo/agriculturalandfoodresearchinitiativeafri.cfm

that match our keywords that notifies us by e-mail. As emphasized in Chap. 14, careful selection of keywords is critical. They need to be narrow enough to screen out numerous items that are not applicable but broad enough to capture those that are relevant.

There are many agencies in the federal government that grant funds for scientific research. The first source for food-science research is the National Institute of Food and Agriculture (NIFA) in the United States Department of Agriculture (USDA). They have several programs, including the Agricultural and Food Research Initiative (AFRI—see Fig. 16.1) and the Specialty Crop Research Initiative (SCRI). Topic areas of interest in the NIFA to food scientists include

• Food Safety & Biosecurity
• Food-Science & Technology
• Hunger & Food Security
• Obesity & Healthy Weight

SCRI emphasizes larger interdisciplinary projects associated with fruits, vegetables, and nuts that span departments, universities, and even regions of the country typically encompassing production, handling, distribution, and economic interests. Many require a sensory component which generally is found in food-science programs. Other federal agencies that fund projects of interest to food scientists are the National Institutes of Health (NIH), National Science Foundation (NSF), United States Agency for International Development (USAID), and the Environmental Protection Agency (EPA).

Another ready source of funding for food scientists is the food industry. Food companies tend to fund narrowly targeted research. Companies in allied fields such as food packaging, ingredient suppliers, and equipment manufacturers, are also interested in the fruits of food-science research. Industry-wide boards, such as the American Cocoa Research Institute and the American Beef Board, grant funds to

A GRANT PROPOSAL FABLE

The USDA had a research program that permitted one proposal from an interdisciplinary research team from each university. There was a very short lead time, but four food scientists put together a proposal combining microbiology, engineering, chemistry and nutrition. They were up against a proposal that represented three nutritionists located in three different departments. The food science proposal was rejected by college administrators because it was not judged to be interdisciplinary since all of the scientists were in the same academic department. It turned out the USDA did not fund the other proposal, but the lesson is that before writing a proposal it is important to know what the evaluators mean by critical terms like interdisciplinary research.

Fig. 16.2 A note of caution when writing a grant proposal

projects that advance industry interests. Another important source of funds for food-science research is a foundation such as the Rockefeller or Ford Foundations.

Once a source has been identified, we need to start working on the application. Completing an application is time consuming. It is a good idea to spend some up-front time to gauge our chances before applying. Carefully review the RFP (request for proposals) on the appropriate website to determine whether we are eligible, how close our ideas fit the goals of the program, deadline date, and success rate (>20% is good). If there are any questions, contacting the program manager of the source is a good idea. We must make sure we are well prepared when contacting the source with a carefully selected set of questions.

As we prepare to apply, we should identify potential collaborators. More on interdisciplinary research was presented Chap. 15. Selection of collaborators should be on the basis of their scientific qualifications. As the strength of collective qualifications of our team increases, our chances of being funded also increase. Some grants reward interdisciplinary projects; others reward projects with collaborators at different institutions (see Fig. 16.2). Remember, however, the more members on our team, the more funds we will have to share with our collaborators if the project is funded. Also, the more departments and institutions involved, the more requirements we will need to meet and the more time it will take. We also need to

- Set a timeline for us and our collaborators
- Learn all the university procedures needed to get the necessary approvals (such as human subjects, animal testing, biosafety, etc.) and signatures
- Learn procedures for patents
- Allot the necessary matching funds from our institution if required
- Determine if we can obtain equipment matching funds
- Learn the ways to cost share when required

When starting the application process, it the responsibility of the principal investigator to make sure that all of the guidelines are followed. Many agencies now ask for a letter of intent to apply. When they get the letter, they make an evaluation as to whether our idea fits their guidelines. The greatest idea in the world must pass by the bureaucrats processing the proposal, only allowing it to be reviewed if it conforms to all of the rules as stated in the guidelines. Federal grant applications are now online. All forms must be properly completed. Typical forms for a USDA project are the cover page, abstract, project description, budget, budget justification, scientist credentials, facilities, certification, conflict of interests, and endorsements. The budget must allow for overhead (the university's cut). For example, overhead for research projects in the College of Agricultural and Environmental Sciences at the University of Georgia is 39%. That means if the proposal limit is $150,000, the most that can be requested is $107,913.67 with the remainder going to the university to provide lights, air conditioning, library services, grant processing, etc. Some granting agencies have limits on overhead amounts (AFRI limit is 22%—reaping $122,950.82 for a request of $150,000), and the university will determine if it will accept those limits.

We can increase our chances for success when preparing the application in several ways. First, we should position our proposal, making sure it meets the program goals and spelling out clearly how it meets those goals. We should take some time to make sure that the proposal fits into the thought collective of our reviewers as that is the basis for their review criteria (even more than the stated criteria). Also, we need to make sure that we emphasize strengths of our proposal and the investigators.

Successful proposal writing characteristics include

- A thorough, up-to-date knowledge of literature
- Familiarity with the prevailing paradigm and how our proposal fits into it
- The latest buzzwords
- Clear and concise writing
- Preliminary data that helps demonstrate the feasibility of our idea
- A track record by the principal investigators
- Adequate facilities and equipment
- A novel idea

We must make sure we know how the application is to be submitted and that it is submitted by the deadline date with all the proper authorizations. Usually it requires authorization by each participating department, the participating college(s), and the university. Do not expect everyone at our university to be available in their offices at 4:30 when the proposal must be submitted by 5:00. It is a good idea to make an appointment with each designated official before crunch time. We need to know what the policies are for getting the proper authorizations. Proposals that cross more than one university require extra signatures and extra time.

After we have submitted the application, there is generally a long wait. Certain proposals, usually long-term ones involving several investigators and millions of dollars will organize a site visit for finalists. Scientists and program staff will come to our location to view the facilities and listen to presentation from each of the collaborators. These site visits need to be carefully planned and choreographed from the time the team arrives at the airport to the time they all depart. They will be observing

- The ability of the team to plan, organize, and execute the visit
- A common vision of team members
- Strong support from the administration

Grant Review Process

The panel style and deliberations vary from agency to agency and evolve over time, but here is some insight into how grant panels operate. The panel chairman reads the abstract of each proposal and selects specific reviewers based on their expertise on the research being proposed. For example, on a certain USDA panel there were 15 members of the panel with a total of 150 proposals submitted. Each of these proposals was read and evaluated by three members of the panel—a primary reviewer, a secondary reviewer, and a reader based on their level of expertise. Each member served as primary reviewer for ten proposals, secondary for another ten, and reader on an additional ten proposals. The groups of three varied from proposal to proposal.

Everything to this point was in writing and done before the panel convened. At the meeting of the USDA panel, each proposal was discussed. The proposal title under discussion, the investigators, and their affiliation(s) were announced. Anyone with a conflict of interest was asked to leave including anyone from the research institution of the proposal investigators. The primary reviewer provided an overview of the project, summarized the comments from the outside reviewers, presented comments, and made a recommendation in one of five categories (High, Medium, or Low priority for funding, Some Merit, or Do Not Fund). The secondary reviewer then added comments and recommendation in one of the five categories. The same procedure was followed for the reader. Questions from other panel members were then directed to the reviewers. If the recommendations from the three who had read the proposal were consistent, the proposal was placed in that category. If they were not consistent, a compromise was found after an open discussion. One proposal generated very heated discussion. The primary reviewer recommended *High Priority* for funding, but the secondary reviewer recommended *Do Not Fund*. The reader was somewhere in between. The proposal was not funded that year, but it was funded by the panel the following year. With 150 proposals, there were many proposals that ended up in each class, except *High Priority* for funding. Thus, proposals within a class were further divided into high, medium, and low priorities within each grouping based on recommendations from the three reviewers of each proposal. The panel members sitting out in the hall for a conflict of interest were then invited back in and the next proposal was discussed. It took four days to complete all 150 proposals.

Once all of the proposals were discussed, the excitement began. All proposals were listed in one of the funding categories. A call went out for any proposals that should be reclassified. After any reclassification was done, then the proposals were ranked in order for all of those that were in the top three categories. Within a category (High, Medium or Low Priority for funding) proposals might bubble up or sink. The panel then was dismissed while the panel manager and the USDA staff person determined who would get funded and who would not. The top-rated projects received full funding. Partial funding was then recommended for middle-rated projects. When the money ran out, a line was drawn. Proposals above the line were

funded, pending the investigators ability to answer questions and make arrange-
ments for funding. Proposals below the line could become eligible for funding if
investigators of funded projects were unable to meet the requirements of the funding agency. No funding was provided for the lower-rated projects. Typical reasons
given for lack of funding were that they

- Were not innovative
- Lacked an adequate knowledge of the relevant literature
- Lacked a clear research focus
- Improperly cited critical references
- Stated objectives that were not readily achievable
- Lacked a practical benefit
- Had investigators not qualified to do the proposed work
- Had inadequate facilities or equipment
- Presented a budget inadequate to achieve objectives
- Were too ambitious

The Way Things Are

Grant writers soon find out that funding is not an ideal world. Like everything else,
the awarding of grants has its own little world and rarely lives up to our idealized
concepts. Here are a few things to think about:

- A good idea is not sufficient in itself to receive funding. It must be presented in
  its best light and in a way to give confidence that it will be executed properly.
- It is not even always the best written proposal that gets funded if it misses a key
  point or is promoting a poor idea.
- There tends to be a bias toward experienced investigators who have a strong
  reputation. This bias may be that they are expected to produce meaningful results,
  that they have mastered the tricks of proposal writing, use the most up-to-date
  buzzwords, or that they are just admired.
- There also tends to be bias against small errors. A good idea paired with a good
  plan may be killed by spelling or grammatical errors.
- Terminology is a key to see who is keeping up with the literature and who is not.
  Obsolete terms suggest obsolete ideas.
- Reviewers are people too and can be appealed to by hopes, fears, and biases. The
  investigator who condemns current thought in the field is probably condemning
  several of the proposal's reviewers.
- The proposal's customers are the reviewers. Positioning the proposal in the context of the current literature offers the greatest chance of success (Ries and Trout,
  2000).
- Proposals too far ahead of their time are likely to be rejected for a low probability
  of success. Reviewers generally reward someone who keeps up and can take it to
  the next logical step.

Table 16.1 Success rates for various grant-funding agency programs (parentheses represent letters of intent)

NSF FY 2008	25%
NSF FY 2009	32%
USDA NRI FY 2006	22%
USDA NRI FY 2007	16%
USDA NRI FY 2008	22% (11%)
USDA AFRI FY 2009	18% (9%)
USDA SBIR Phase I	11%
USDA SCRI FY 2008	31% (9%)
USDA SCRI FY 2009	40% (13%)

Sources for information:
http://www.csrees.usda.gov/funding/nri/nri_annual_reports.html; http://www.csrees.usda.gov/fo/sbir.cfm
http://www.agnr.umd.edu/news/images/2010%20NCERA-101%20
Report.pptx; http://www.sciencemag.org/cgi/reprint/326/5957/1181.pdf;
Sheely et al., 2010

- Divergers are the investigators most likely to be rewarded because reviewers and agencies tend to favor problem solvers.
- In viewing the résumés, reviewers tend to be more impressed with research publications than review articles and book chapters. Participation in other grants is also viewed favorably as long as they are followed up by timely publication of results.

Breaking into the class of funded scientists is a very difficult assignment with rather low success rates (Table 16.1). For every funded proposal, there are at least two and maybe as many as six proposals that are not funded. That amount can go up to nine or ten letters of intent that are not funded for every one that is funded. In addition, experienced grant writers are likely to have a higher rate of success than those who have never been funded. Despite the odds, every successful grant writer broke through for the first time once. There are books that will provide some insight into the grant writing process (Blackburn, 2003; Chapin, 2004) and websites (http://www.federalgrants.com/grant-writers.html; http://www.science-funding.com/about.html), but grant programs vary so much from program to program that it is difficult to provide recommendations that apply to specific programs. Many programs provide webinars on grant opportunities and national meetings may offer insights into funding decisions.

Two ways to learn how to write effective grants are to find a mentor or write as many proposals as it takes to be successful. Possible mentors are our major professor while still in school or a colleague when we start seeking our own grants. Our major professor may be more than happy to help us learn the craft but may also view us as competitors, if not now then in the future. Likewise, a colleague might view us as competitors. Volunteering to be a lead investigator with a colleague for programs they do not usually apply for might be a way to get a good start. When our grant application is rejected we should not be satisfied with just the written comments. Communication with the panel director will help us gauge whether it is worthwhile to revise and resubmit. Chances are the panel director will not give us a direct yes-or-no answer but will probably help give some direction if we are good at reading between the lines.

Managing a Grant

The odds of having a grant funded may be against us, but some applicants actually are funded. Notification of our grant being funded is a cause for celebration, but it is also a call for action. When we receive funding from an agency or company, we need to start the planning process for implementation if we have not already, determine how the funds will be distributed, develop ways of managing the grant, and develop a timeline for timely report writing as required by the granting authority. We may need to readjust the budget if the award is less than the amount requested. If our grant is only partially funded, which objective(s), etc. get sacrificed? How much can we do? How will this partial funding affect our co-investigator(s)? Now is the time to negotiate with our collaborator(s) and the funding agency. Read the notification letter carefully. Is there anything else we need to do? For example, have we received human subjects or recombinant DNA approvals pending? What must we do to get funds? What are the deadlines?

Grant management is another skill we must master or future grants will be even more difficult to obtain. Ramping up from no funding to one grant can be a challenge to a newcomer. With the pressure to obtain grants, investigators may submit several proposals within a calendar year. The worst-case scenario is no funding. The next worst-case scenario is more funding than we can handle effectively.

When grant funds come into a university, they usually go to a specific department and it is sometimes difficult to work out the logistics of spending money across two or more departments. In addition, it is sometimes difficult to properly apportion credit. If a research team is small, all investigators tend to receive reasonably equal credit for the innovative nature of the team and numerous publications. As the team matures, however, and the membership changes, newer members may be regarded as providing service for the original members and not providing leadership, a key component of a scientist's reputation (Rosei and Johnston, 2006).

Effectively managing a grant generally requires that we effectively manage the funds we receive and the new people we will need to hire to get the work done. We are also responsible for proper scheduling of activities and events. Three years may seem like a long time, but it will probably take us at least six months to ramp up to speed. Time rushes by, particularly if we are dealing with perishable, seasonable crops. We will also need to make sure to meet all the reporting requirements of the grant. In a federal grant, periodic reports must be written. We can expect visits from representatives of the company for industry-funded grants. Other responsibilities include making sure we publish our results in a timely manner if the grant is publicly funded and ensuring that any proprietary research is kept confidential from all parties except the funding party. Our institution also has specific requirements that we must meet. Finally, we may need some extra time to complete our mission. When this happens we can request a no-cost extension, which most agencies will honor. This extension will allow us to wrap up the research for this grant and expend the funds remaining that were not spent during the original period of the grant (Blackburn, 2003; Chapin, 2004).

Receiving funding does not necessarily reduce pressure. Usually it increases pressure. Once we are among the funded scientists, we now have personnel including postdocs, graduate students, and technical staff who are dependent on our support. The larger our research group becomes and the more publications we generate, the easier it is to generate preliminary data for the next proposal and the greater our reputation will be for successful research. As a result, our success rate for funding is likely to increase, but funding is not guaranteed. Loss of expected funding requires effective management. Many universities provide some bridge funds to established researchers who may have a temporary gap in funding to help keep the lab going. Some graduate schools also have funds to support graduate students complete their degrees if their major professor has lost funding.

References

Blackburn TR (2003) Getting science grants: effective strategies for funding success. Jossey-Bass, San Francisco, CA
Chapin PG (2004) Research projects and research proposals: a guide for scientists seeking funding. Cambridge University Press, Cambridge, UK
Ries A, Trout J (2000) Positioning; The battle for your mind. McGraw-Hill, New York
Rosei F, Johnston T (2006) Survival skills for scientists. Imperial College Press, London
Sheely D, Poth M, Jerkins D (2010) AFRI 2009 Annual Synopsis, USDA

Chapter 17
Laboratory Setup and Management

Power is getting people or groups to do something they don't want to do.

Leslie Gelb

Setting up a Laboratory

To this point we have worked in a laboratory, probably our major professor's. Soon we will probably become the lab manager. It looks easy when an underling. It becomes a much bigger deal when it is all of our responsibility. We most likely will be taking over an existing laboratory. This transition is the easiest one in the short-term, but it is frequently the most constricting long-term. We will definitely want to add some of our personality to the lab. We may be involved in renovating an existing laboratory. This opportunity will delay our ability to get up and running quickly, but it will provide us with more flexibility later if we plan well. Some time in our career, we may have the opportunity to design a new laboratory. It is a great experience if we know what we want, but it can become a real drag on our time during the process. Finally, we may be involved in designing a new laboratory as part of the design of a new building. Again, a new building is exciting but very time consuming. Detailed descriptions of laboratory design and all the things we need to consider are provided by Dahan (2000) and DiBerardinis et al. (2001).

Regardless as to how we acquire a laboratory, there are many functional considerations we need to make before getting started. The laboratory must be organized to meet our primary research or analytical focus. What do we want to achieve during the first one-to-five years in the lab? Also, we need to consider a secondary focus. The primary mission may not always work out and we need a viable Plan B or we may be looking for another job. In a university or government setting, we may also want to build in a little flexibility in the lab to allow for a shift in future research directions. To keep the lab in a university going funding support is critical. Research priorities in industry and government can change rapidly. Successful labs keep just

R.L. Shewfelt, *Becoming a Food Scientist: To Graduate School and Beyond*,
DOI 10.1007/978-1-4614-3299-9_17, © Springer Science+Business Media New York 2012

ahead of those changes. Finally, there may be nonresearch functions that need to be considered in a lab such as desk space, etc.

We may have the opportunity to purchase some equipment. This time is not the time to be shy or to put it off. There will probably never be a better opportunity to obtain high-ticket items than when we have just been hired. Usually there is an equipment package that goes with the position, but it will not be available forever. We may only have a few weeks or a few months to spend the available funds. In a university, we probably will be given a figure of how much to spend. In industry, there may be money in our supervisor's budget that we will not get if we do not ask for it. Focusing on getting the most for our money is critical. Some things we need to consider when putting together our wish list include what

- Needs to be accomplished in the lab during the first year and over the longer term
- Equipment available now and how functional it is
- Equipment will provide the biggest impact in the first evaluation period
- Will provide the best long-term investment

Also, do not overlook the potential of leveraging funds with a collaborator to get a bigger piece of equipment that neither can afford on our own.

There are also logistical considerations that we must make. Most labs are either organized by territory or by work station. Territorial assignment of benches to different workers in the lab is effective when there is more space than workers and when there is little overlap between projects. Work stations designed to perform a specific type of test, usually designed around a specific instrument or prep work, become necessary when there are more workers than space and there is more overlap in projects. Most labs work in a mixed mode with every worker having some space to call their own within the context of work stations for frequently used procedures. In addition, traffic flow patterns and congestion must be considered. We should design the lab to minimize interference between workers. Also, do not forget the laboratory environment. Instruments must be kept away from dust and in areas that do not fluctuate widely in temperature. There must be proper storage areas for chemicals, chemical waste, and supplies that meet university or company guidelines. Finally, we want an environment that is conducive to productivity and a good attitude among laboratory personnel. An example of a well-organized laboratory is shown in Fig. 17.1.

Food scientists may also have the occasion to design, operate, and manage pilot plant facilities. Pilot plants provide the opportunity to manufacture small batches of food products using miniaturized equipment that mimic manufacturing operations. They are important in scaling up new product prototypes from bench-top formulations in industry or to test process variables in industrial or academic research. Such facilities require special attention, particularly with regard to cleanliness, sanitation, and safety. An example of a university pilot plant is shown in Fig. 17.2.

In all of these laboratory settings, we will also be responsible for laboratory staffing. Although we may inherit the worker(s) when we first take over a lab, we will eventually be hiring and perhaps terminating employees. Every person in the lab will probably be assigned to one or more projects. In addition, they might have specific roles

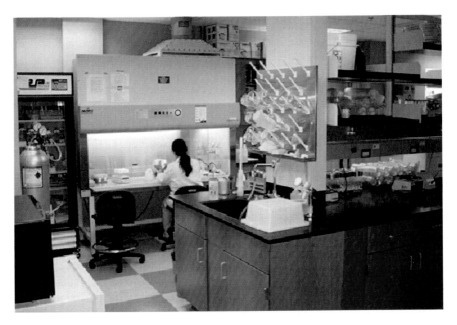

Fig. 17.1 Modern food microbiology laboratory. Photo courtesy of Dr. Mark Harrison

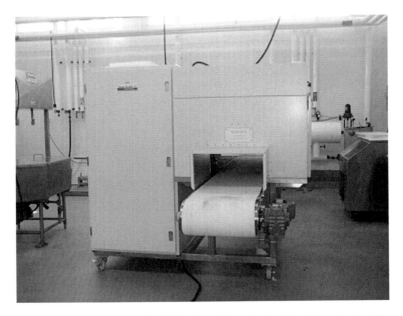

Fig. 17.2 Radiofrequency oven in the University of Georgia pilot plant. Photo by Sara Yang

such as ordering supplies, monitoring chemical waste, equipment maintenance, etc. When bringing a new person in the lab, we must carefully evaluate their qualifications to perform specific tasks and their potential interactions with others already in the lab. Labs take on their own personalities, frequently an extension of the personality of the lab manager. Compatibility between people in the lab is important.

We also must distinguish between permanent and temporary employees. For example, a major professor may have a lab run by a technician or postdoc. The technician may be full-time or shared with other professors. Generally speaking, technicians tend to be more permanent than postdocs and graduate students. It is critical that the relationship be good between the lab manager and the permanent employee(s) as the manager's eyes and ears in the lab, the protection of equipment, and the training of temporary employees will probably all fall on the permanent staff. If there is tension between a manager and permanent employee(s), it must be resolved either by making peace or reassignment. A festering relationship greatly decreases any chance of success in a lab.

Permanent employees provide continuity in the lab that a temporary employee cannot, but the major contributions to our research reputation and funding will be through our students and postdocs. Developing the proper balance between permanent and transitory employees is difficult. Graduate assistantships tend to be the most cost-effective way to generate large amounts of research for an Assistant Professor who can attract motivated students and provide the necessary training and mentorship. As a research program grows, some professors prefer to employ postdocs who are able to focus their full attention on research without the distractions of courses and exams. Mature labs with strong funding are typically run by one to three postdocs supported by several graduate students and a technician or two. Graduate students are attracted by scientific reputation. Recruiting good graduate students by researchers just starting out is facilitated by interaction with undergraduate students through courses, Food Science Club activities, undergraduate research in our lab, and IFT Student Division activities in regional sections and the national meeting. Postdocs are recruited by following the literature and attending presentations of doctoral students at regional and national meetings. Marketing the lab via a well-designed and up-to-date website is also effective.

Managing the Laboratory

Laboratory personnel have specific needs. It is important that they all receive proper training for the tasks they need to perform. Training may come from formal courses offered by the company or university, from different people in the laboratory or from others outside the lab. Everyone in the lab should have clear assignments, adequate tools to do the job, adequate space in the lab, and enough freedom to be creative. In addition, we need to make sure there is a mechanism for communication within the lab and with the lab manager.

Other issues that we must face as the lab manager include staff morale. If the manager is not happy, neither is anyone in the lab. In *Power Rules,* Gelb (2009) argues that a leader needs to be both liked and feared to be effective. Achieving the proper balance is critical. Pushing too hard breeds resentment and probably reduced productivity. Not pushing hard enough limits our ability to achieve our goals. We should recognize that each person in the lab has goals and ambitions of their own. If our goals are aligned with theirs, chances are that both will be met.

One of the most difficult tasks of any manager is dealing with personal problems and conflict resolution within the lab. The ability to listen is more important than the ability to speak when dealing with individuals and conflicts between individuals. The lab manager is also primarily responsible with the interface of the lab with other laboratories. Many times, more progress is made between labs when it does not involve the lab managers directly. If interaction between labs can be more informal and confined to the lead assistants in each lab, misunderstandings can be minimized. Ultimately we are responsible, however, for what happens in our lab or how personnel in our lab interact with people from other labs.

Within the lab, the lab manager wields the power, but each manager must realize that there are constraints. The institution, university/governmental agency/company, has guidelines for acceptable and unacceptable laboratory setups and personnel policies. Always make sure that the lab is following the rules. Economic constraints limit the type of equipment we can provide, the numbers of workers in the lab, and the supplies we can purchase. Space limits the types of tests we can run and the number of workers we can hire, although we might be able to schedule people in shifts to increase lab productivity. The qualifications and productivity of lab personnel limit what can be accomplished by the lab. Likewise, available equipment limits what we can and cannot do. Remember that the capability of our lab can be extended by collaboration with colleagues who are willing to share the equipment in their labs. We must make sure that personnel in our lab are using equipment in our lab and other labs responsibly or we will have big bills to pay and loss of access. Finally the biggest constraint of all is time—ours and that of everyone in the lab. For a detailed description of everything a lab manager is likely to face and more, see Barker (2010).

To facilitate communication within our lab and with those who may be interested in our lab, we should consider setting up a website for the lab (Barker, 2010). Be sure that anything put on that site complies with organizational rules and guidelines. Few things can be as effective at marketing a program as an impressive website, but an out-of-date site may be worse than none. Bestowing a name on our lab as shown in Fig. 17.3 give it a sense of identity. Periodic face-to-face meetings are essential for an effective laboratory. Many labs have weekly sessions, but a critical mass is necessary. When there are just a few members within the lab, one-on-one sessions may be more appropriate. When critical mass has been achieved, typically one member will present an update on their project or lead a discussion on a recent publication with relevance to the group. Such sessions provide students with experience in presenting their data in nonthreatening situations and foster an *esprit de corps*. When there are large numbers involved, the lab may be subdivided into subgroups arranged around research areas. Occasional joint meetings between subgroups or with other lab groups can add interest to these sessions. Such meetings can generate enthusiasm for research, but they should not become busywork.

Fig. 17.3 Cartoon by Sidney Harris. Reprinted by permission of ScienceCartoonsPlus.com

Some simple guidelines for setting up a laboratory are as follows:

- Organize the lab around the primary function making sure that it receives the bulk of available resources
- Build in flexibility in case priorities shift
- Consider the costs and benefits of sharing with other labs
- Before making major changes, learn reasons for existing conditions
- Improve efficiency in context of maintaining effectiveness
- Recognize that a laboratory is a dynamic entity and that we need to spend time in it to keep in touch with the changes going on
- Develop an organizational scheme that embodies clarity of responsibility, a viable mechanism of communication, instrument and personnel scheduling, and problem resolution

References

Barker K (2010) At the helm: leading your laboratory, 2nd edn. Cold Spring Harbor Laboratory Press, Cold Spring Harbor, New York

Dahan F (2000) Laboratories: a guide to master planning, programming, procurement and design. W.W. Norton & Co., New York

DiBerardinis LJ, Baum JS, First M, Gatwood GT, Seth AK (2001) Guidelines for laboratory design: health and safety considerations, 3rd edn. Wiley, New York

Gelb L (2009) Power rules: how common sense can rescue American foreign policy. Harper, New York

Chapter 18
Career Development

> *The most common commodity in this country is unrealized potential.*
>
> Calvin Coolidge

It may be hard to focus on a career when still in school, but it is never too early to start developing a long-range plan. Developing a career is more than just seeking a job. This chapter will start with job seeking and then put it in the context of a career plan as was discussed in Chap. 15.

Seeking a Job

The first step in any search for a job is self-evaluation. Some questions to consider include:

- What is our career goal?
- What is our objective for the next 5–10 years? Will accomplishment of this objective help you get to our career goal?
- What is our primary objective for the next job? Will accomplishment of that objective help us achieve our objective for the next 5–10 years and our career goal?
- Is our current job moving you toward your long-term objective and career goal?

If the answer to any of these questions is no, we need to either re-evaluate our plans, objectives or career goal.

Next, we should establish priorities for our next job. There are several considerations. The most frustrated people in life are those who try to have it all and end up with no satisfaction! What are the top five priorities in Fig. 18.1 in order of preference? See the job description in Fig. 18.2. How well does it fit our aspirations? How does it match up with the priorities set in Fig. 18.1?

R.L. Shewfelt, *Becoming a Food Scientist: To Graduate School and Beyond*, DOI 10.1007/978-1-4614-3299-9_18, © Springer Science+Business Media New York 2012

_____ setting (government, academe, industry)
_____ type of job within that setting
    _____ government (regulatory, research, administration)
    _____ academic (teaching, research, extension)
    _____ industry (R&D, QC, management, technical sales, regulatory, etc.)
_____ geographical location,
_____ salary,
_____ work hours,
_____ work environment,
_____ opportunity for advancement,
_____ fringe benefits and vacations,
_____ job security,
_____ compatibility with your supervisor,
_____ coworkers that are easy to work with, and
_____ institutional philosophy.

Fig. 18.1 Considerations for choosing a job. Rank them in priority as they meet your needs

GRAIN STATE UNIVERSITY

ASSISTANT PROFESSOR OF FOOD MICROBIOLOGY

Outstanding individual sought for tenure-track assistant professor position in the Department of Food Technology. Successful candidate is expected to produce a grant-funded program of research excellence. Grain State emphasizes team learning and partnership with industry. This 12-month research and teaching position is available for Fall Semester. The holder of this position will be expected to teach undergraduate courses in Food Microbiology and Food Fermentations and develop a specialty course to be taught at the graduate level. Experience with distance education is preferred.

Salary is competitive accompanied by an excellent benefits program. A **Ph.D. in Food Science**, microbiology, genetics or related discipline with emphasis on food pathogens is required. Effective written and oral communication skills are critical.

Submit credentials, college transcripts, publication abstracts, statement of research and teaching interests, and letters from at least three references by May 15 to:

    Dr. William V. Oser, Search Chair
    Grain State University
    Department of Food Technology
    1939 Willey Way
    Borlaug, GR 99962-1939

Grain State University is an Equal Opportunity/Affirmative Action Employer

Fig. 18.2 Mythical job advertisement

Once our priorities have been established and on how the job we seek fits into our long-range goals and objectives, it is time to design a base résumé. Remember that the primary purpose is an introduction to land an interview. There are many questions we should ask:

- Does it present a professional appearance?
- Is it too cluttered?
- Is it too sparse?
- Is it in a clear, crisp style?
- Is it too long or too short? For an industry job, limit it to one page. For an academic job, one and a half pages are fine plus any presentations and publications.

- Does it emphasize the strongest attributes first?
- How does it measure up on content?
- Does it have enough personal data for them to contact us?
- Does it have more personal data than we want them to know?
- Does the e-mail address reflect a serious professional or cutesy college kid?
- What is the objective? Career Centers like to have us put our objective on it. Be careful as it may tend to eliminate more jobs than it does to help.
- Does it provide the necessary information on education? At minimum, it should probably have the degree (or anticipated degree) and date, GPA out of 4.0, thesis and/or dissertation titles.
- Does it highlight work experience with particular emphasis on experiences in food science and or management etc.? Do not forget significant projects completed in school.
- Does it list the most important honors, awards, and special skills with an emphasis on those related to food science? I was a star newspaper carrier in the ninth grade, but I usually leave that out.
- Does it list the most significant extracurricular activities, particularly those that show professional affiliation or leadership?
- If it is for an academic position does it list your presentations and publications? Give complete citations?
- Does it list references? Unless specifically called for, leave these off. When providing references, make sure that at least half of them are professional references. References are usually much more important for academic or government positions than ones in industry.

Chris has designed a résumé which is shown in Fig. 18.3. It has some strong points, but it also has weaknesses and omissions. Can you spot them? For more details on making the most out of a résumé consult a good book on the subject such as Whitcomb (2010) or a university career guide (University of Georgia, 2011).

A résumé is usually sent with an accompanying cover letter. Even if the cover letter is sent via e-mail, it should be more formal and professional than an ordinary e-mail. From the cover letter and the résumé

- Will they be able to find what they are looking for?
- Does it demonstrate an ability to translate experience into results?
- Does it show a general competence for the position and a breadth of background?
- Does it reflect a positive attitude?
- Does it project self-confidence? Make sure the attitude does not come across as arrogant or shy.

Chris is sending his résumé to apply for the position at Grain State. His cover letter is shown in Fig. 18.4. Has he met all the requirements for a strong cover letter?

CHRIS APPERT
canner@milku.edu

EXPERIENCE
Caddy – University Golf Course (2009 to present)
Carried clubs for touring pros.
Served drinks and snacks to dignitaries.
Some maintenance of greens.
Wait staff – Village Green Restaurant (August, 2005–July, 2009)
Took orders and served food and beverages to clientele.
Served as cashier when necessary.
Helped develop unique menu items.

HONORS AND AWARDS
Harvey W. Wiley Fellowship, $25,000/year, (2009-present)
IFT Scholarships in Freshman (2005/6), Junior (2007/8) and Senior (2008/9) years
Presidential Scholar, Summer 2006, Spring 2008, Dean's List, Fall 2006, Spring 2007, Fall 2007
National Merit Scholar (2004/5)
Megafoods Product Development Contest, team leader, 2nd place (2010)

EDUCATION
Ph.D.   Milk University. expected May, 2013. Food Science. GPA 3.87.
           Dissertation Title: Quantitative Exploration of Blueberry Volatiles and Their Role in Consumer Acceptability
BS   University of Meat and Beans. May, 2009. Biology. Magna Cum Laude. GPA 3.92.

RELEVANT CLASSROOM EXPERIENCES
FDST 6010/L Served as QA Manager and Corporate Coordinator in a virtual food company.
   •   responsible for developing all quality tests for four products.
   •   was the chief management officer in the company and transmitted all class reports to the instructor.
FDST 6030/L Self-directed study on the isolation and diagnostics for *Edwardsiella tarda*.
FDST 6080/L Familiarized with HPLC/GC/LC, ICP-OES, and ELISA methods.
FDST 6110/L Team leader, Blueberry Fresh Pack, micro-perforated HDPE package to extend shelf life.
FDST 6250/L Team leader, Bacon and Egg Rolls, bacon and egg filling in a bite-size egg roll.
FDST 6320/L HACCP certification
   •   designed HACCP plan for value-added turkey product
FDST 8020/L Team leader on report "Salt Consumption and Reduction in the American Diet".
   •   surveyed over 200 undergraduate students on salt-use and consumption.
   •   conducted sensory-descriptive and consumer tests on the effects of salt reduction in peanut butter, luncheon meats and baked products.
   •   submitted abstract for presentation at IFT and manuscript for publication in Journal of Food Science Education.

SKILLS
Proficient with Microsoft Word, Microsoft PowerPoint, Microsoft Excel, Harvard graphics, SAS.
In-depth experience with GC/MS, GC Olfactometry, MDGC, FRAP analysis, SDS-PAGE and aseptic techniques.
Conducted Sensory Descriptive Analysis (panel leader), difference tests, consumer tests, and focus-group interviews (panel moderator).

EXTRACURRICULAR ACTIVITIES
Milk University Golf Intramural Club (2009-present).
PETA chapter, Milk University, secretary-treasurer (2010/11).
SALSA dance club (2010-present).

PERSONAL INTERESTS
Biking, Golf, SALSA dancing.

REFERENCES
Available upon request.

Fig. 18.3   Mythical résumé for a Ph.D. student ready to graduate. Do you see any potential problems with this résumé?

The Interview

A successful cover letter and résumé leads to an interview. We should have at least one or two questions for them before even going to the interview. The last question usually asked by the interviewers is "Do you have any questions for us?" An absence of questions is frequently interpreted as a lack of interest in the position. When I was interviewing at two universities, I asked the question "If I needed to buy $100 worth of chemicals (back then $100 was real money!) how would I go about it?" I received clear, consistent answers from several people at one university but not from the other location. The response was a factor in my decision to come to the university that still employs me more than thirty years later. When asking these

Dr. William V. Oser, Search Chair
Grain State University
Department of Food Technology
1939 Willey Way
Borlaug, GR 99962-1939

Dear Dr. Oser,

Please consider my application for Assistant Professor of Food Microbiology as advertised in *Food Technology*. My research is nearing completion, and I expect to graduate in May. I would be ready to start this summer in preparation for teaching in Fall Semester. Enclosed are my résumé and transcript.

Although I had planned to work in the food industry, the opportunity to teach at Grain State University is just too tempting to dismiss. I enjoy teaching and have been a Teaching Assistant in three courses here at Milk University. I believe that it is very important to engage students as they learn. Several of my classes have involved experiential learning including an exercise in Food Microbiology involved in the development of diagnostics for an assigned pathogen. It was the most meaningful educational experience that I have ever had in college. I intend to bring active learning into the classroom and to use current literature as a basis for a graduate-level course in Food Bioinformatics.

I am confident that I can develop a grant-based research program in bioinformatics. I have presented three posters at IFT annual meetings and published two manuscripts from my research. A third manuscript has been submitted for publication. I have been fortunate to have the mentorship by my major professor, Dr. Babcock, on grant writing. She asked me to help her write the grant proposal on blueberry volatiles recently funded by AFRI. I fully understand the pressures on junior faculty to produce, and I am prepared to meet the challenge.

I firmly believe that I have the qualifications to be a credit to Grain State in this position. Fee free to contact me if you require additional information.

Sincerely,

Chris Appert
Harvey W. Wiley Fellow
Milk University

Fig. 18.4 Mythical cover letter for an advertisement in Fig.18.3. Do you see anything wrong with this letter?

questions, we should make sure we do not come across as one who is more interested in benefits, vacation, and perks than work, and also we do not come across as a nerd. Usually the best employees are ones that are hard workers who are effective on the job and who fit into the institutional culture and the assigned working group.

After the interview, expect a letter in the mail, a phone call, or even an e-mail. Make sure the message on our phone is professional. A formal letter is usually an indication that we were not selected for the position. Everyone who has ever applied for a job has probably been rejected. Do not take it personally. I was turned down to work at McDonald's once. I was crushed, but it was not the end of the world, and I still eat a Big Mac from time to time. If possible, try to get some feedback from the interview. It is time to get in touch with a personal contact made during the interview. Ask this person for indications on what went well and what we could have done better. Let the person talk, but do not press them because we may be putting them in a difficult position with the way lawsuits go these days. Also, do not say anything negative about the process or the organization. My wife had an interview with a small college, and it was clear that they had already selected someone else who had better qualifications than she did. Several weeks later, a few days before classes started, she got a call offering the position to her because the chosen candidate backed out at the last moment. That became my wife's first full-time professional position, and she held it for three years until we moved away. Any rejection is time to consider reevaluating your approach to interviews, but do not try to be somebody who you are not. It may work in the short run, but it can lead to a job that does not suit you.

A phone call is usually the way an offer is made followed up with an e-mail attachment or fax. Regardless of how we feel about the company, remain calm. It is usually best to indicate that we will get back to them, even if we are ready to jump at the chance. Look carefully at the written offer if they provide one. Assess the advantages and disadvantages of that job. If we have had other interviews and are interested in another offer, get on the phone, indicate that we have another offer, and let them know we need an answer by a certain date. Be careful in playing one company against another. I have known graduates who have been very successful in increasing their initial salary offer, but there is always the possibility that the company we want to work for will wish us well working for the other company. IFT provides periodic salary information (see Kuhn, 2010).

Developing a Career

As in the search for a job, it is even more important to evaluate ourselves in developing a career plan. Carefully consider a career goal, a 5–10 year objective, and job objective. In addition to professional considerations think about how important are

- Ambition (How important is it to become rich and famous?)
- Home/social life (Do we even need one? What are we willing to give up?)
- Idealism (What happens when ambition runs into conscience?)
- Likes/dislikes (Are we willing to perform unpleasant tasks to get ahead?)
- Strengths/weaknesses (How can we best exploit strengths and overcome weaknesses?)

Academic Careers

There are many career options for food scientists (see Hartel and Klawitter, 2008). To this point, we have probably been exposed to more food scientists involved in academic institutions than in any other setting. Professors teach, conduct research, and perform outreach (also known as Cooperative Extension). In universities, we can also move into administration, but only after having proven ourselves in at least two of the main areas (teaching, research, and outreach). Teaching is what most students think with respect to professors, but it is the research that is most important in making the reputation of a professor. It is critical that we are able to bring in funds to our program if we are going to survive in the academic arena as discussed in Chap. 17. A less stressful role with a much lower salary in universities for food scientists is as a technician.

In developing a career plan, keep in mind typical career patterns within your chosen direction. In universities, a member of the faculty proceeds from Assistant Professor to Associate Professor to Professor and maybe to endowed chair. The promotion and tenure system is known as up-or-out or publish-or-perish. Typically after five years, the Assistant Professor must demonstrate a national/international reputation among peers as demonstrated in a dossier of accomplishments including research publications in top-tier journals and an ability to attract grant funding. The dossier also includes about five letters of evaluation by experts in the field of inquiry. Successful candidates are promoted to Associate Professor and granted tenure. An unsuccessful candidate must look elsewhere to continue to be employed. A more rigorous process is conducted about five years later for promotion to Professor, but unsuccessful candidates are still retained by the university. Although mobility is possible from one university to another, most university personnel make their career at one location. The primary movers from one university to another are research superstars or those who make a move to become a Department Head. Dr. Lloyd Walker used his success in the lab and classroom to move into university administration as shown in Fig. 18.5.

Government Careers

Another avenue for the food scientist is in federal, state, or local government or allied nongovernmental agencies. Regulatory agencies like FDA, USDA, EPA, and state Departments of Agriculture need food scientists to write, interpret, and enforce regulations. They also need scientists with the capability of conducting analyses or developing new methods of detections. Many federal agencies conduct basic and applied research without the obligations of teaching or outreach. Also there are management opportunities in these organizations.

Government has specific grades based on qualifications. They do not have tenure as such, but government employees are subjected to a rigorous evaluation process during promotion and can be terminated if they are not promoted and do not conform to specific benchmarks. Dr. Barbara Schneeman, an example of a food scientist who went into government after a highly successful career in academia, is profiled in Fig. 18.6.

Dr. Lloyd Walker
Alabama A&M University

Alabama A&M University (AAMU), Huntsville, Alabama
        Interim Associate Provost with Responsibilities for
                Institutional Research, Planning and Sponsored Programs
        Interim Associate Provost for Academic Affairs and Dean, University College
        Professor and Chair, Department of Food and Animal Sciences

Education
        BS, Animal Science, Prairie View A&M
        MS, Animal Science, Prairie View A&M
        PhD, Texas A&M
        Postdoctoral training, Food Chemistry/Biochemistry, Alabama A& M

Honors include:
        Who's Who Among America's Teachers, 1998, 2004, 2006, 2009
        Excellence in Teaching Award, 2000(AAMU)
        Researcher of the Month, 12/2000, 12/2001, 10/2003 (AAMU)

IFT Activities:
`   Phi Tau Sigma - Member
        Chair, Dixie Section   (1998 -1999)
        Chair, South Eastern Section (2007 – 2008)
        Member, National Diversity Committee (2004 – 2005)

Publications:
        More than 150 Abstracts
        More than 45 refereed papers
        Three (3) book chapters

Fig. 18.5 Career Profile of Dr. Lloyd Walker

Dr. Barbara O. Schneeman
Director: Office of Nutrition, Labeling, and Dietary Supplements
Center for Food Safety and Applied Nutrition
Food and Drug Administration

University of California, Davis
    Professor, Departments of Nutrition, Food Science and Technology,
    Internal Medicine
    Head, Department of Nutrition
    Dean, College of Agricultural and Environmental Sciences
    Associate Provost for University Outreach
University California, San Francisco (Sabbatical leave)
    Visiting Scientist Cardiovascular Research Institute

Education
    BS, Food Science & Technology, University of California
    PhD, Nutrition, University of California, Berkeley
    Postdoctoral training, gastro-intestinal physiology, Children's Hospital,
    Oakland

Honors include:
    Fellow, American Association for the Advancement of Science
    Carl Fellers Award, Institute of Food Technologists
    Harvey W. Wiley Medal, Food and Drug Administration
    Samuel Cate Prescott Award, Institute of Food Technologists

Fig. 18.6 Career Profile of Dr. Barbara Schneeman

Careers in Industry

The principal employer of food scientists is the food industry as well as allied companies. Food companies need knowledgeable personnel to manage quality control/assurance operations. The most glamorous aspect of the food business is product development. Most seasoned developers have a portfolio of products they designed. It is this aspect of food science that I stress in recruiting new students to the major. Other opportunities in the industry include technical sales, package development, process engineering, plant production, and management. Allied industries include ingredient suppliers, packaging companies, trade publications, and analytical laboratories.

In general, the food industry has more lucrative salaries with less job security than academe or government. Promotions in industry are generally achieved by replacing someone higher up in the chain, either your boss, someone in another department at the same location, or by moving to another location. Movement from one company to another is common in the food industry. Food companies are also likely to go through mergers and acquisitions which generally mean that some R&D people will be released to gain cost savings. Obtaining an MBA, usually through night or weekend programs, is recommended. Most companies will pay for an MBA with a commitment to stay with the company for a given length of time. An MBA with a technical background is highly valued in food companies and usually rewarded with a healthy increase in salary. Profiles of Dr. Raghu Kandala and Dr. Gillian Dagan are provided in Figs. 18.7 and 18.8 as examples of careers in R&D and an analytical-laboratory company.

Other Opportunities

A fourth area of opportunity is to become self-employed such as becoming a consultant to the food industry or starting your own company. It is usually best to gain some experience before going this route. Some people like the freedom of being their own boss, but most prefer the security of a regular paycheck which is not guaranteed to consultants. There is not a clear career plan for the self-employed. I would recommend much varied work experience in the food industry and mentorship by someone in the field. China Reed, profiled in Fig. 18.9, has used a background in the food industry to provide a basis for a consulting company she started. Some brave souls work across the different types of career patterns in food science. Experience in one area (academe, government, industry, self-employed) is usually beneficial in another area, but there are many cultural differences between the areas. It is very difficult to master the politics in one type of system and then adapt to a very different type of system where the rewards and expectations are different.

Dr. Raghu Kandala
Frito-Lay R&D Manager

Frito Lay, Plano, Texas
    Research Scientist (2005-07);
    Principal Scientist (2007-2011)
    R&D Manager (2011 – Present)

Nestle PTC, Beauvais, France
    Intern (07/00 – 03/01)
    Researched novelty ice cream products

Education
  B.Tech (Hons), Agricultural Engineering, IIT, Kharagpur, India
    M.S., Agricultural and Biological Engineering, Penn State
    Ph.D., Food Science and Technology, UGA

Honors include:
    Frito Lay Quality Award
    Frito Lay Team Award
    IFT Graduate Scholarship
    JJ Powers Graduate Scholarship.

Contributions at Frito Lay:
    Developed and launched 4 dips in the marketplace.
    Developed process solutions and thermal processes for various existing and new
    products.

Fig. 18.7 Career Profile of Dr. Raghu Kandala

Dr. Gillian Dagan
Chief Scientific Officer, Product Performance Services
ABC Research, Gainesville FL

Education
       BS, 2000, Food Science and Human Nutrition, University of Florida
       Ph.D., 2004, Food Science, University of Florida

Certifications and Skills
       QDA and Sensory Spectrum Trained
       Meat and Poultry HACCP certified
       Instrumentation: GC-FID, HPLC, UV-VIS, Electronic Nose

Honors and Service
       Graduate Research Paper Competition, IFT Nonthermal Division, 2004
       Proctor and Gamble, Phi Tau Sigma Award for Excellence in Scientific Research, 2004
       Chair, Florida Section of IFT, 2006-2007
       University of Florida, College of Agriculture Alumni Horizon Award, 2010

Fig. 18.8 Career Profile of Dr. Gillian Dagan

China A. Reed
Senior Food Scientist/ Product Developer
Uniquely Yours Consulting

Education
       B.S. Food Science, University of Illinois
       M.S., Food Science, University of Georgia

Employment
       Food Technologist, Best Foods/ Unilever, 1998-1999
       Senior Food Technologist, Schwan Food, 2000-2004
       Founder, Uniquely Yours, 2004-present

Contributions
       Developed In Store Bakery Lemon Burst Meringue Pie, Krogers
       Developed Mardi Gras Cheesecake, Popeye's
       Developed Snickers Cheesecake, Bob Evans
       Clients include Georgia Spice Company and Kantner Group

Fig. 18.9 Career Profile of China Reed

References

Hartel RW, Klawitter CP (2008) Careers in food science: from undergraduate to professional. Springer, New York

Kuhn ME (2010) 2009 IFT membership employment & salary survey. Food Technology 64(2): 20–37, http://members.ift.org/NR/rdonlyres/80DA4BD5-033B-40C6-9B62-CC47FA98996B/0/0210feat_salarysurvey.pdf

University of Georgia (2011) Career Guides. Available from http://career.uga.edu/resources/career_guides/

Whitcomb SB (2010) Résumé magic: trade secrets of a professional résumé writer, 4th edn. JIST Works, Indianapolis, IN

Index

Printed by Publishers' Graphics LLC